Cloning
A Beginner's Guide

D1146432

From anarchism to artificial intelligence and genetics to global terrorism, **BEGINNERS GUIDES** equip readers with the tools to fully understand the most challenging and important debates of our age. Written by experts in a clear and accessible style, books in this series are substantial enough to be thorough but compact enough to be read by anyone wanting to know more about the world they live in.

anarchism
ruth kinna

anti-capitalism
simon tormey

artificial intelligence
blay whitby

biodiversity
john spicer

bioterror & biowarfare
malcolm dando

the brain
a. al-chalabi, m. r. turner & r. s. delamont

christianity
keith ward

cloning
aaron d. levine

criminal psychology
ray bull *et al.*

democracy
david beetham

energy
vaclav smil

evolution
burton s. guttman

evolutionary psychology
r. dunbar, l.barrett & j. lycett

fair trade
jacqueline decarlo

genetics
a. griffiths, b.guttman, d. suzuki & t. cullis

global terrorism
leonard weinberg

hinduism
klaus k. klostermaier

life in the universe
lewis dartnell

the mafia & organized crime
james o. finckenauer

NATO
jennifer medcalf

the palestine–israeli conflict
dan cohn-sherbok & dawoud el-alami

philosophy of mind
edward feser

postmodernism
kevin hart

quantum physics
alastair i. m. rae

religion
martin forward

the small arms trade
m. schroeder, r. stohl & d. smith

FORTHCOMING:

animal behaviour

beat generation

bioethics

british politics

censorship

christianity

climate change

conspiracy theories

crimes against humanity

engineering

ethics

existentialism

extrasolar planets

feminist theory

forensic science

galaxies

gender & sexuality

globalization

hinduism

human rights

humanism

immigration

indigenous peoples

modern slavery

oil

philosophy of religion

political philosophy

racism

radical philosophy

renaissance art

romanticism

socialism

time

volcanoes

Cloning
A Beginner's Guide

Aaron D. Levine

ONEWORLD

OXFORD

CLONING

A Oneworld Book
Published by Oneworld Publications 2007

ISBN-13:978–1–85168–522–6

Typeset by Jayvee, Trivandrum, India
Cover design by Two Associates
Printed and bound by TJ International, Padstow, Cornwall

Oneworld Publications
185 Banbury Road
Oxford OX2 7AR
England
www.oneworld-publications.com

Contents

Preface

This book aims to introduce the reader to the fascinating science of cloning. It assumes no prior knowledge and seeks to highlight key advances and ongoing research in plain language. Fortunately, while challenging in practice, the general cloning procedure is simple in concept. And knowledge of this general technique is sufficient to understand both the immense promise of cloning science and the controversy research in this field creates.

Cloning came to the forefront of the world's stage when the birth of Dolly, the famous cloned sheep, was announced in 1997. This book recounts the history leading to this startling event, which changed the field of cloning science forever. But the focus is more recent advances. In the last decade, cloning science has advanced rapidly, opening new doors and posing new risks.

Our focus will be three important categories of cloning research. The first of these, animal cloning, is the most developed. Although controversial in some quarters, it is poised to impact the food we eat, the medicines we take and numerous other facets of our daily lives.

The second category, human therapeutic cloning, is less developed and more controversial. This research is closely linked with work on human embryonic stem cells, which we will also examine. Scientists believe these master cells, which are isolated from human embryos, may, in the long run, prove useful for treating a variety of diseases. In the still-unproven therapeutic cloning protocol, scientists create cloned human embryos from which they can isolate these medically promising cells. The use of cloned embryos in the protocol theoretically should lead to embryonic stem cells that are genetically matched with individual patients, greatly reducing the risk of immune rejection during therapy.

While animal cloning and human therapeutic cloning are pursued by mainstream scientists around the world – at least in countries where the research is deemed acceptable – the third category, human reproductive cloning, has been pushed to the fringes of the scientific community. Still, some people claim to have attempted to produce cloned human children and ongoing research, with other goals in mind, makes reproductive cloning attempts more likely to succeed in the future. Indeed, the possibility that a cloned human will one day be born looms large over the entire field of cloning science.

The ongoing and unresolved debate over the ethics of cloning research and the impact of this debate is another focus. Bioethicists have argued about cloning for many years. Only recently, however, since mammalian cloning became a reality and since the isolation of human embryonic stem cells opened the door to the possibility of human therapeutic cloning, have these arguments had such an impact. Indeed, ethical considerations have led to restrictions on research, or the funding available for research, in the United States and numerous other countries. These restrictions are shaping how much research is done and where it is undertaken. They may, in the long run, also influence the development of medical therapies, particularly those using the therapeutic cloning protocol.

Throughout the book, I have sought to approach these controversies in a balanced manner, giving equal weight to ethical arguments for and against a particular use of cloning technology and examining both the potential of and the challenges facing the related science. My personal beliefs, no doubt, have slipped through in a few places, but my goal has not been to persuade, but rather to allow readers to reach their own conclusions.

Cloning science stands poised to shape our future, both in exciting and in potentially frightening ways. But just what these specific impacts will be remains an open question. Debates taking place today, over the regulation and the legality of cloning and related embryonic stem cell research, will influence cloning's future. A key goal of mine in writing this book has been to create a resource that will help readers come to their own conclusions about cloning technology and participate meaningfully in the debates shaping its, and our, future.

Aaron D. Levine
2007

Acknowledgements

I owe a debt of gratitude to many people for their varied roles bringing this book into existence and eventually to completion. Lee Silver, who first suggested I undertake this project and who commented on many key portions of the text deserves special thanks. I also owe thanks to the staff at Oneworld Publications. I am grateful to Juliet Mabey, Marsha Filion, and others for trusting me with this project and for their assistance throughout the process. Marsha, who provided detailed comments on an initial draft of the manuscript and arranged for an anonymous reader to do the same, played a crucial role shaping the final manuscript.

I am also grateful to the many colleagues, friends, and family who took the time to review sections of the manuscript, engage in helpful discussions or simply provide encouragement. While space permits me to list only a few, I would like to extend my thanks particularly to Harold Shapiro, whose guidance during the proposal stage shaped the final product, and Derek Chiang, whose comments and insightful suggestions influenced key sections of the book. Finally my mother, who read every word of the book, often several times, played a crucial role helping me ensure the cutting-edge science I describe remained accessible to a non-scientific reader. Above all, I want to thank Amy, who graciously put up with the many late nights and early mornings this book required and whose encouragement throughout helped bring the project to completion.

While I am grateful to all who helped shape the book, I, of course, chose to follow some suggestions and ignore others. I also made difficult decisions regarding what material to include and what to leave out. Accordingly, I take full responsibility for the book's various shortcomings.

List of illustrations

1

What cloning is and why it matters

Cloning technology was invented during the twentieth century and now is poised to help define the twenty-first. Almost everyone has heard of Dolly, the cloned sheep born in 1996 but what about the rapid progress made since then? Scientists now count horses, cows, cats, and dogs among the many animals they can clone. This progress raises a host of questions. Are you comfortable drinking milk or eating meat from a cloned cow? Should we clone extinct or endangered species? Will the April 2005 birth of Snuppy, the world's first cloned dog, usher in a new era of cloned pets? Should we clone embryos to generate embryonic stem cells and help develop medical therapies? And perhaps the most important question of all: when, if ever, will this progress lead to the first cloned human?

Although scientists are nearly unified in their opposition to cloning humans for reproductive purposes, on-going research toward other goals makes this likely, if not inevitable. For the most part, this research is driven by the hope that cloning technology will have significant health benefits, perhaps leading to transplantation therapies that use embryonic stem cells specifically tailored to individual patients. Of course, if a cloned human is ever born, the desire for fame will almost certainly play a role. Looking back to the media frenzy surrounding the birth of the first test tube baby in 1978 or the clamor surrounding the birth of Dolly, it is not hard to imagine the furor that a cloned human baby would generate.

As modern biotechnology is increasingly applied to humans, it raises important questions for society to address. Should we, perhaps in the relatively near future, allow infertile couples or single mothers to use cloning technology to try to produce a child? Should we, in the longer term, permit parents to use cloning technology not just to have children, but to have children with specific genetic modifications or enhancements? Debates on cloning technology and its implications are, all too often, hijacked by advocates or opponents who skew the science to fit a particular view. Although the details of cloning research are complex, the general technique is not particularly difficult to understand. And understanding this general technique and its consequences is more than enough to participate fully in these important debates and to see through the many myths clouding discussions of cloning.

What cloning is

Cloning is, at its most basic level, reproduction without sex. "Sex" does not refer to the act of intercourse but to sexual reproduction – the joining of genetic material from two parents into an embryo that may, if development goes well, give rise to a new adult organism. All humans alive today were born through sexual reproduction; a single sperm from the male joined with an egg from the female, creating an embryo with half its genetic material derived from each parent. This mixing of genetic material introduces an element of chance into reproduction, ensuring that children differ genetically from their parents. In cloning, offspring are genetically identical to their single parent. Such offspring are the products of "asexual" reproduction.

Cloning, rather than relying on the merging of egg and sperm, uses the genetic material or DNA from a single cell. This cell is joined to an egg from which the DNA has been removed. Next, this construct is coaxed to develop as if it were a newly fertilized egg. If development proceeds normally, the resulting organism will be genetically identical to the single donor. In this case, reproduction no longer generates new combinations of genetic material but faithfully duplicates previously existing ones.

Although mammals do not normally reproduce asexually, nature does provide a close analogy: identical twins. Roughly one out of every 250 human births results in identical twins – siblings that are genetically identical. Because a cloned child would be genetically identical to its DNA donor, it can be helpful to imagine cloning as a form of delayed twinning. If cloning technology were perfected and applied to humans, the birth of a cloned human would not be altogether unlike the birth of identical twins but instead of a few minutes separating the two births, there could be many years.

Scientists speculate that a cloned human and his or her parent would typically be less similar than identical twins. This is because the environment plays an important role in development. Identical twins usually share much of the same environment, while a cloned human and his or her genetic parent often would not. Identical twins develop in the same uterus and usually grow up in the same household. In contrast, a cloned human would probably be carried in a different womb and grow up in a different household from its genetic parent. The cloned child would also be born into a world that had changed significantly. The importance of environmental influences has led bioethicists who have considered the possibility of human cloning to focus on its unpredictability. It is not clear that a child cloned from Mozart or Pavarotti would grow up to perform or even appreciate music.

Humans have not been cloned and few plausible reasons exist to clone humans for reproductive purposes. Some have suggested that cloning might provide a means for infertile parents to have a genetically related child. However, fertility research seems likely to lead to other, more effective and less controversial, approaches to treat the few couples for whom this last resort might be necessary. Others have suggested cloning may be justified when a child dies young; believing parents would deserve a chance to bring their lost loved one back to life. But many think this would lead to

THE NATURE VS. NURTURE DEBATE

The human cloning debate is closely intertwined with the long-running argument over the relative importance of genes (nature) and the environment (nurture) in shaping people's physical attributes and beliefs. This debate pits those who believe one's place in life is determined primarily by innate abilities against those who see cultural and environmental influences as the most important factors shaping human lives.

Numerous attempts have been made to tease out the relative importance of these two factors, including comparisons between identical and fraternal twins. The story that has emerged is complex. Few traits are purely genetically or environmentally determined: most are influenced by both factors. Height, weight, intelligence, and many personality traits, to name just a few examples, are shaped by the interplay between one's genes and one's upbringing. However, the relative roles played by these two factors are unknown, ensuring that the nature vs. nurture debate will live on to be argued another day.

disappointment all round. Due to environmental influences, the cloned child would not be the same as the deceased child he or she was ostensibly replacing. Furthermore, the new child, forever competing against an idealized memory, might face unreasonable expectations. In the end, neither parents nor child would prosper.

Because human cloning seems remote and is generally undesired, cloning science today focuses primarily on animal research. In animals bred for human use, such as cows, pigs, and horses, the advantages of asexual reproduction are significant. The element of chance central to sexual reproduction frustrates animal breeders and livestock producers. When mating a prize-winning stud to promising mare, horse breeders aren't excited by the chance that the resulting foal will randomly receive the parents' worst genes: they want to propagate the genes that turned the stud into a champion, in the hope of producing future winners. Cloning, by allowing breeders to produce genetic replicas of valuable animals, makes this process more efficient. For horse racing, this efficiency comes at a steep price, as cloned horses are currently forbidden from participating in officially sanctioned races. These sorts of restrictions don't apply to pigs or cows, which are bred to produce meat and milk for consumers, rather than for competition. Not surprisingly, livestock breeders, particularly in the United States, have shown interest in using this technology to make their operations more productive and more profitable.

What cloning is not

By and large, cloning is not what you see in the movies. It is not photocopying; or at best it is like using a slow and blurry photocopier – so slow, that by the time the copy is made, the original has changed. If you cloned your dog today, there wouldn't be an exact replica running around and barking tomorrow, as suggested in the Arnold Schwarzenegger hit, *The Sixth Day*. Rather, you would create an embryo that could potentially be transferred into the womb of a surrogate mother. Nine weeks later, if all went well, a puppy would be born. This puppy would be genetically identical to your dog but, obviously, much younger. It might look like its parent had looked as a puppy but it would experience a different environment and, perhaps, mature differently.

Movies such as *Multiplicity*, in which an overworked contractor clones himself to help cope with his busy life, ignore the time delay essential to cloning. In this case, the movie's premise, while entertaining, is absolutely wrong. The clones, rather than helping out at work and around the house, would be a burden. They would be infants, not adults as portrayed in the movie, and like any human infants would need nearly constant attention. As any parent can tell you, adding a baby (or several) to your family is not a good strategy for gaining extra time.

Nor does cloning bring back the long-dead. Cloning technology, at least at its current efficiency levels, requires a significant amount of biological material. For living animals, it is simple to take a sample and preserve this material: Dolly, for instance, was cloned from frozen cells. However, finding enough genetic material presents a significant hurdle to cloning long-extinct species. For now, the cloning of dinosaurs, as seen in *Jurassic Park* and its successors, is no more than a scientific pipe dream. That said, scientists have made progress in cloning endangered species and some believe cloning may offer a promising conservation strategy. Attempts to clone recently extinct animals, such as the Tasmanian tiger, where preserved biological material may still exist, remain a possibility.

As we shall see, cloning is not easy. When Dolly was born, she was the only success in 277 attempts. Success rates have improved but the procedure remains inefficient. Many cloned embryos fail to develop, and when development does start, a variety of abnormalities are seen. Even in the most efficient operations, only a minority of the original cloned embryos develop to term and go on to lead healthy lives. At the moment, this inefficiency limits the usefulness of animal cloning for commercial purposes. It also raises the ire of animal rights activists, who complain that the technology produces deformed animals. Obviously, these inefficiencies would need to be overcome before scientists could even begin to consider cloning humans for reproductive purposes.

Why cloning matters

Cloning matters because it is on the verge of affecting daily life around the world and its importance will only grow with time. Animal cloning will revolutionize food production in the coming years and may, by turning animals into biological factories, revolutionize pharmaceutical production as well. Moving from animals to humans, cloning technology may, if some expectations prove true, radically alter medicine, leading the way to an era of personalized transplant therapies. Finally, in the longer term, it opens the door to the cloning (and potential genetic engineering) of humans, perhaps changing the very essence of what it means to be a human being.

A growing scientific consensus suggests that milk and meat from cloned animals, or at least from their progeny, are safe for human consumption. In December 2006, the U.S. Food and Drug Administration announced preliminary plans to allow products from cloned livestock into the food supply. If finalized, such a ruling could have dramatic effects. Scientists can clone several important farm animals, including cows and pigs, but only a small number of cloned animals – none destined for consumption – live on American farms today. One industry insider has estimated that within twenty months of a ruling allowing products from cloned animals into the food supply, American farms would be covered with hundreds of thousands of clones.[1] This could occur despite widespread consumer discomfort with the very idea of eating products from cloned animals.

Thus far, the United Kingdom and most other European countries have shown more caution regarding the introduction of cloned animal products into the food supply. If, as appears likely, the United States approves these products first, it could contribute to continued trade wars. Although cloning does not necessarily include genetic modification, some cloned products will almost certainly also be genetically modified. Thus, trade in cloned products could get tangled in the on-going debate on the import of genetically modified organisms; a number of countries have limited their imports of agricultural products from nations where genetic modification is prevalent.

When Dolly was cloned in 1996, the research was primarily funded by a biotechnology firm that aimed to revolutionize the way drugs are produced. We'll learn more about this later but the basic idea is to create, through cloning, genetically modified sheep or cows that produce therapeutic compounds, such as insulin or growth hormone, in their milk. Pharmaceutical companies could isolate these valuable compounds from the milk for a fraction of the cost of traditional manufacturing methods. The milk would not be intended for human consumption and would probably be discarded after the therapeutics had been isolated. This technique, known as "pharming," offers potential economic benefits for drug companies and has taken off since Dolly's birth. Numerous cows have been bred to produce therapeutics in their milk and some scientists are exploring the possibility of harvesting drugs from other body fluids, including urine. Pharming raises a number of concerns, including the risk of drug-producing animals accidentally entering the food supply. Although the risks may be remote, even those of us unfazed by drinking milk from a cloned cow wouldn't be pleased to find out the milk was significantly enriched with a prescription medicine.

While cloned animals that produce therapeutic compounds already exist, the creation of cloned human embryos to facilitate medical therapies remains in the future and raises serious ethical questions. Many scientists are optimistic that cloning will, one day, regularly be used to create stem cells genetically matched to specific patients. These cells could, potentially, help treat a range of debilitating conditions, such as type 1 diabetes and Parkinson's disease. Because the cells would be genetically matched to the individual patient, they might avoid the immune rejection problems that complicate transplant therapies today. This potential therapeutic technique is controversial, however, because deriving these patient-matched stem cells, using currently envisioned approaches, would require the creation of a cloned human embryo. At five days of age, the stem cells would be isolated from the embryo and the developmental process halted. Dramatic advances toward this vision of regenerative medicine were reported by a group of researchers based in South Korea, but in late 2005 the veracity of this work was called into question: today, it is clear that most, if not all, these advances were fraudulent. Despite this set-back, many scientists believe the vision remains promising and "therapeutic cloning" is being pursued by scientists around the world.

Cloning also matters because, given the field's current trajectory, it is part of our shared future. From the food supply to the medicine cabinet, cloning technology is poised to change the way we live. But these changes are controversial. Each of us can and should participate in the debates that will shape the role cloning plays in the future. Before you say "yuck" to drinking milk from cloned cows or rush off to save your dog's DNA in preparation for eventual cloning, take the time to learn a bit about the science. Although cloning is fairly simple, misinformation is prevalent. Understanding the science behind cloning will help make these debates more meaningful and their outcomes more satisfactory for everyone.

2

A cloning parts list: cells, genes, and embryos

In some ways, cloning is remarkably simple. The process can be described in just a few words: scientists (in the early twenty-first century) start with a healthy unfertilized egg and an adult cell. They remove the genetic material from the egg and replace it with the genetic material from the adult cell. They then trick this reconstructed embryo into developing as if it were a newly fertilized egg. If all goes well, this cloned embryo is transferred into the womb of a surrogate mother and develops normally.

This simple description raises many questions. What is the genetic material inside a cell? Where is it located? Will any adult cell work for cloning or is a specific type required? What is it about an unfertilized egg that allows it to re-program an adult nucleus and lead to normal development? What is normal embryonic development and how can you tell if a cloned embryo is developing normally?

To answer these and other questions, this chapter briefly surveys some of the biology underlying cloning; providing a parts list to understand the technology. Cloning science incorporates insights gained by biologists working in a wide range of fields and, as we build this parts list, we will review a number of important biological concepts, including heredity, DNA, cells, and mammalian development, and see how important discoveries in these areas paved the way for cloning.

This overview begins with heredity, the study of the transmission of characteristics from one generation to the next. Although "like begets like" is a truism dating from ancient times, it was only recently, in the twentieth century, that scientists started to understand the mechanism of heredity, or how junior ends up with his father's jaw and his mother's curly brown hair, not to mention grandpop's knack for numbers and grandma's not-so-reliable memory. This understanding, incomplete as it is, relies on the identification of deoxyribonucleic acid, DNA, as the genetic material and on the understanding of DNA that scientists have developed since its structure was first described in 1953. Just as identical twins are identical because they share a DNA sequence, clones are clones because they share a complete (technically, a nearly complete) set of DNA. For this reason, we will examine the basics of DNA, focusing on its structure and how its sequence codes for proteins.

Although the discovery of DNA's role in heredity was a key step in the development of cloning technology, understanding DNA is not enough. What really matters is the relationship between DNA and cells, the building blocks of life. The smallest organisms consist of just a single cell, while humans are made up of countless trillions. Almost without exception, every cell contains a full complement of an organism's DNA. Furthermore, when a cell replicates and divides, its DNA, in a carefully choreographed dance, replicates and divides as well. Many of the advancements in cloning technology, in particular, the cloning of Dolly in the late 1990s, relied on an appreciation of the importance of this intricate process and, for this reason, this chapter will briefly introduce cellular structure and division.

One cell divides into two. Two into four. Four into eight. And so on. Through this process a single egg, whether fertilized by sperm or created with cloning technology, develops into a multicellular organism and eventually into an embryo, fetus, and new individual. We will introduce this process with a brief look at mammalian development, highlighting key events and the various stages relevant to cloning technology.

Mendel's garden and the laws of inheritance

Parents and children often look, and even act, alike. This was no doubt as true three thousand years ago as it is today, yet it belies easy explanation. Why do some children resemble their mothers, while others look more like their fathers? Why do some children not really resemble either parent? These and similar questions have captivated scientists and non-scientists alike throughout recorded history.

Aristotle, writing in the third century BCE, outlined a theory of heredity, based primarily on observation, which proved remarkably prescient. He recognized that children often displayed characteristics of both their mother and their father, which implied that both parents must contribute some sort of material during reproduction. Although this conclusion seems obvious today, it was not so then. The belief that men were the sole providers of genetic material, while women served merely as incubators, persisted until the eighteenth century. Aristotle also noted that characteristics acquired later in life, such as the loss of an arm by a soldier in battle, were not passed on to children. From these and other observations, Aristotle proposed a theory of inheritance that relied on some sort of non-material information passing from parents to their children. Science went into a steep decline following the fall of Rome and Aristotle's theories, although they were not always either well-known or accepted, remained state-of-the-art well into the sixteenth century.

Although many contributions were made to the science of heredity in the 1600s and 1700s, we will turn directly to the work of Gregor Mendel in the 1800s. Mendel, a monk living in Brunn,

Austria (now Brno, Czech Republic), was the first person to study heredity quantitatively and so revolutionized the field. "Quantitative" means "counting" but this seemingly simple step represented a significant advance. When Aristotle concluded that both parents contributed to their child's genetic inheritance, because some children resembled their mothers and others resembled their fathers, he was relying on qualitative observations. In contrast, Mendel, who studied peas rather than humans, counted how many of his plants had white flowers and how many had purple flowers. These numbers gave Mendel a key insight into the actual mechanism of heredity.

Mendel was a clever, diligent, and careful experimenter. His decision to work with peas was the first of many wise choices. Peas were an ideal organism for the study of heredity for a number of reasons. A key benefit was that a large number of visually distinct varieties of peas existed. The clear distinctions between pea plants with purple flowers and those with white flowers or between plants with smooth seeds and those with crinkled seeds, to cite just two examples, made it easy for Mendel to identify the results of his experiments.

The pea varieties were also true-breeding. This simply means that their characteristics (flower color, seed shape, etc.) remained the same generation after generation. If Mendel planted a plot of purple-flowered peas and let them reproduce on their own, through self-fertilization, all the resulting offspring had purple flowers. While peas will self-fertilize if allowed, it is easy to block this process and manually fertilize the plants. This technique was central to Mendel's work, as it allowed him to cross true-breeding purple-flowered plants with true-breeding white-flowered plants and record the results.

To understand Mendel's experiments and their importance, let us turn our attention to a simple cross between a true-breeding purple-flowered plant and a true-breeding white-flowered plant. According to the dominant theory of the day, known as

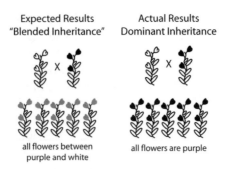

Figure 1 Results of Mendel's cross between white-flowered and purple-flowered pea plants. The dominant theory of the time suggested that all offspring would be intermediate in color between the two parents. However, Mendel observed that all offspring had purple flowers and concluded that purple was a dominant trait.

"blended inheritance," this cross should have yielded flowers that were intermediate in color between purple and white. Instead, Mendel observed that all the offspring had purple flowers. This result held true for a large number of crosses: for example, when Mendel crossed pea plants with yellow seeds with plants with green seeds, he found that all of the offspring had yellow seeds.

Although the striking uniformity of the first generation (which Mendel described as the "dominance" of one trait over another) represented a significant advance, Mendel didn't stop at the first generation. He allowed this generation to self-fertilize and observed the results. In his flower color experiment, he found that when the first generation of purple-flowered plants was allowed to self-fertilize, the second generation were mostly purple-flowered but with a smattering of white-flowered plants: the purple-flowered plants produced by his first cross were not true-breeding. This relationship held true for yellow and green seeds, round and crinkled seeds and many other traits. While the first generation was uniform in appearance, with one trait dominating, in the second generation Mendel observed both the original traits. The dominant trait was always more common than the other trait, which he called "recessive." This is how Mendel's counting proved its value: while other scientists had probably seen similar results, Mendel was able to identify a pattern. For each of the crosses he studied, Mendel noted that the ratio of the dominant characteristic to the recessive one was roughly 3:1.

Mendel continued his experiments by investigating how the second generation of plants passed on their characteristics to future generations. He found that all the recessive plants (for example, the white-flowered plants) of the second generation were true-breeding but the purple-flowered plants were less consistent. Some were true-breeding, while others produced both purple- and white-flowered offspring. After careful tabulations, Mendel concluded that the 3:1 ratio he observed in the second generation was really a disguised 1:2:1 ratio. In other words, in the second generation of his flower color experiment, 25 percent of plants were true-breeding purple, 50 percent were non-true-breeding purple and 25 percent were true-breeding white.

Based on these findings, Mendel developed a model of inheritance that, tweaked and modified, remains the foundation of modern genetics. Mendel proposed that parents do not transmit actual physiological traits to their offspring but rather discrete information about these traits. Mendel called this information "factors." He also proposed that an organism received two factors for each trait, one from each parent. In some cases, the factors might be identical and code for the same characteristic; in other cases, the two factors might be different. When the factors were different, only one, the dominant trait, was expressed; the recessive trait was not visible. Mendel also proposed that the maternal and paternal versions of these factors were indistinguishable. It didn't matter if the dominant trait was inherited from the mother or the father; it was expressed in either case. The factors also remained "uncontaminated." Even if a recessive trait was not expressed for several generations, it persisted unchanged and could be expressed in a later generation. From another, more complex, experiment, Mendel concluded that the factors were independent. Thus, inheritance of a factor coding for flower color proceeded independently of inheritance of a factor coding for seed shape or plant height.

Mendel's results, and his model, were striking but his work was published in a rather obscure journal and remained unknown for almost thirty-five years. When it was rediscovered, in 1900, it quickly revolutionized the study of heredity. Mendel's theory was particularly remarkable given his lack of knowledge of the inner workings of cells and the structure of the factors he envisioned.

Today these "factors" that contribute to heritable traits are called genes. We know that a gene is a segment of DNA that codes for a particular protein. Furthermore, genes are packaged in chromosomes carried in egg and sperm cells and transmitted from parents to children during reproduction. Mendel knew none of this.

While Mendel's model remains useful, many important human traits are influenced by multiple genes and inherited in a more complex manner. For example, genes play a role in determining both human height and intelligence, yet these traits are influenced not by one but by tens or hundreds of genes. However, simple Mendelian inheritance can explain some important human conditions, including cystic fibrosis, a recessive genetic disease that affects one out of every two to three thousand Caucasian newborns. Approximately eight million Americans are genetic carriers of cystic fibrosis and, as Mendel's theory predicts, when two of these carriers reproduce, approximately one quarter of their babies are born with the disease. Huntington's disease is another human disease that follows simple Mendelian inheritance rules but unlike cystic fibrosis, Huntington's disease is dominant. Thus, it is expressed (in the same way purple flowers were expressed) whenever the variant of the gene that causes Huntington's disease is inherited from either parent. As Mendel's theory predicts, approximately half the children born to a parent with Huntington's disease will receive the affected gene and eventually develop the disease.

Mendel's theory of inheritance and, in particular, his proposal of discrete factors responsible for transmitting traits from one generation to another provided an early impetus for the research that eventually led to cloning. To understand how the field developed, we will move on to DNA and learn more about the genetic material that Mendel concluded must exist but about which he knew nothing.

Chromosomes and the search for Mendel's factors

Advances in microscope technology, around the time of Mendel's work, allowed scientists to begin to untangle the inner structure of cells. We'll learn more about cell structure later but what matters for now is that in the late 1800s, scientists begin to observe chromosomes: long thread-like structures in the center of cells. Once Mendel's laws were rediscovered, it wasn't long before scientists realized that these structures had a lot in common with his proposed factors.

In 1902, Walter Sutton, then a graduate student at Columbia University, first noted this similarity. Through his work on grasshopper cells, he noticed that most cells had two copies of each chromosome but that egg and sperm cells had only one copy. He also realized that during *meiosis* (the process by which sperm and eggs are formed) two copies of each chromosome paired and then separated into different cells. This pairing and separating provided a potential mechanism for the process Mendel had proposed. Indeed, if different factors were found on individual chromosomes, it could explain how a purple-flowered plant might give rise to both purple- and white-flowered offspring. Theodor Boveri, a professor

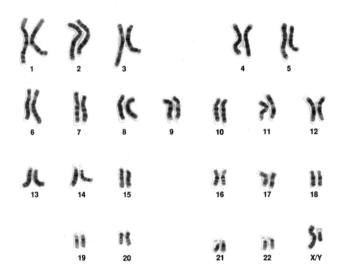

Figure 2 A karyotype showing the chromosomes from a single human cell. There are two copies each of chromosomes 1 through 22 and one copy each of the X and Y chromosomes. (Image courtesy of the Talking Glossary of Genetics)

at the University of Würzburg, came independently to the same conclusion and their theory eventually became known as the Boveri-Sutton chromosomal theory of inheritance. This theory was controversial and left open as many questions as it answered. Over time, however, it gained acceptance, particularly following the discoveries that sex was determined by chromosomes and that other traits were linked to sex-determination and thus could only be inherited from a single parent.

The identification of chromosomes as the carriers of genetic information was a critical step in the development of an understanding of heredity. Yet, while chromosomes provided an explanation of the transmission of information from one generation to another, they provided little insight into what this information was or how it acted within the cell. The identification of chromosomes did not even explicitly identify the genetic material. Rather it set off a scientific controversy because chromosomes consist both of proteins and DNA and for many years it was unclear which of these carried the genetic information.

Several elegant experiments in the first half of the twentieth century resolved this debate in favor of DNA but a key mystery remained: how could DNA, such a seemingly simple molecule, contain so much information? It was known that DNA was primarily made up of four different sub-units (called nucleotides): adenine (A), cytosine (C), guanine (G), and thymine (T); what was not clear was how these four "bases" (as these nucleotide building blocks are typically called) could contain the amount of information necessary to explain heredity. The answer to this question relied on understanding the structure of DNA and so its discovery marked a turning point in the study of biology. DNA's structure was so elegant that its elucidation led quickly to an understanding of the genetic code and to the development of a model for DNA replication.

The DNA double helix

The structure of DNA was determined in 1953, when James Watson and Francis Crick proposed that the molecule was a double helix – a spiral of two interwoven strands of nucleotide bases – and that genetic information was determined by the sequence of the bases. Each strand contained the same sequence but the two strands ran in opposite directions. This permitted complementary pairing of bases, between the two strands, which held the helix together. In Watson and Crick's model, A paired only with T and C paired only with G. This pairing, by weak interactions (known as hydrogen bonds) explained reported but not understood data on the distribution of the bases in an organism's DNA.

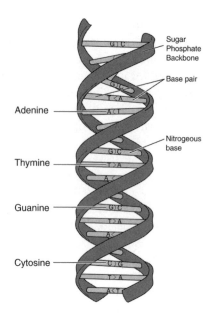

Figure 3 The DNA double helix. The four bases are arrayed in order along a sugar phosphate backbone. Complementary base pairing between the two stands can be seen. Adenine pairs only with thymine and cytosine pairs only with guanine. (Image courtesy of the Talking Glossary of Genetics)

Hereditable material needs, by definition, to be passed accurately from one generation to the next. One of the appeals of the double helix model was that it suggested an obvious means for DNA replication. This was not lost on Watson and Crick, who ended their short 1953 paper describing DNA's structure with the oft-quoted understatement "it has not escaped our notice that the specific pairing we have postulated immediately suggests a possible copying mechanism for the genetic material."[1] This mechanism involved breaking the hydrogen bonds holding the two strands together and using each strand as a template, again based on complementary base pairing, for a new DNA molecule.

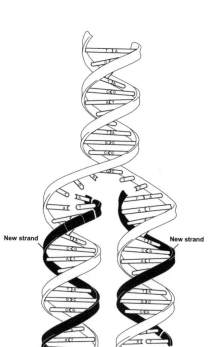

Figure 4 A model for DNA replication. Exact replicas of DNA molecules are created when the original double helix unwinds and the exposed bases serve as templates for the creation of new molecules. (Image courtesy of the Talking Glossary of Genetics)

Unlike Mendel's findings, Watson and Crick's DNA paper was published in one of the world's pre-eminent science journals and was immediately recognized as a groundbreaking advance. Follow-up experiments quickly verified key elements of the model and some of its predictions, including the model of DNA replication it suggested. Other work focused on deciphering the genetic code: the process of determining how the sequence of bases within the DNA molecule contained useful information. It is no exaggeration to say that Watson and Crick's discovery was a critical step in the creation of entirely new sub-fields of biology and the development of the multi-billion dollar biotechnology industry.

As we shall see, one of the questions that drove the development of cloning technology was the question of how, if at all, the genetic material in an organism changed during development. Watson and Crick's identification of DNA as the genetic material proved crucial to this. Today, the tables are turned. We know that DNA doesn't normally change during development but cloning is seen as a useful strategy to engineer deliberate changes in an organism's DNA.

Unraveling the genetic code

While the description of DNA's structure was an important advance, cloning scientists are particularly concerned with how DNA exerts its influence within the cell. This is important because, while DNA, for example, is the heritable material, its sole function is the storage and transmission of information. It doesn't really work within the cell: DNA codes for blue eyes but doesn't turn eyes blue. It is proteins that do the work of the cell. In eye color, genes (of which at least three exist) code for proteins involved in the production of a pigment called melanin and the amount of this pigment in the iris of the eye determines if eyes appear brown, blue, or some other color.

To understand how DNA codes for eye color, let's look at what scientists have learned about the relationship between DNA and proteins. Remember that DNA is a component of chromosomes. Although different species may have different numbers of chromosomes, all members of a species have the same number of chromosomes in their cells. Human cells have 46 chromosomes, two copies of the 22 chromosomes found in both males and females

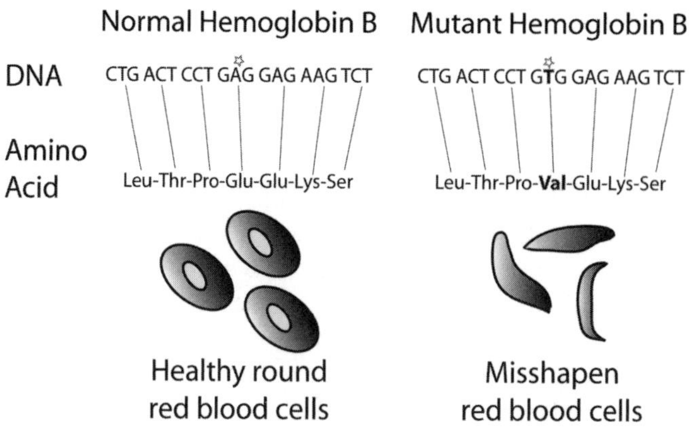

Normal Hemoglobin B   Mutant Hemoglobin B

DNA   CTG ACT CCT GĂG GAG AAG TCT   CTG ACT CCT GṪG GAG AAG TCT

Amino
Acid   Leu-Thr-Pro-Glu-Glu-Lys-Ser   Leu-Thr-Pro-**Val**-Glu-Lys-Ser

Healthy round   Misshapen
red blood cells   red blood cells

Figure 5 The genetic code and sickle cell disease. A single change in the DNA molecule (from A to T) alters the amino acid sequence and ultimately the functioning of red blood cells. The misshapen or "sickled" red blood cells have a harder time moving through small blood vessels and the resulting lack of blood can cause pain, infection, or damage to vital organs.

and two sex chromosomes. There are many more heritable characteristics than there are chromosomes and so each chromosome must contain many genes. The DNA molecule that forms the core of each chromosome contains hundreds, if not thousands of genes, and these genes are, for the most part, distinct segments of DNA. A gene may be as short as a thousand bases or, in rare cases, as long as a million. For eye color, scientists have identified three relevant genes, two on chromosome 15 and one on chromosome 19 and suspect that several more play a part.

Scientists know that most genes code for proteins, complex molecules that do most of the work of the cell. (In the example of eye color, the proteins of interest are enzymes – substances that help chemical reactions proceed more quickly.) Proteins are built up from twenty amino acids; the DNA sequence of a gene specifies the order of amino acids within the protein. Each amino acid is specified by three nucleotides: this correspondence between nucleotides and amino acids is the "genetic code." Because each cell can have thousands of copies of a given protein, small changes in a single DNA molecule can lead to huge differences within the cell and ultimately in human health. For most people with cystic fibrosis, the difference between health and illness is three missing bases and for people with sickle cell anemia, an inherited disease in which red blood cells are misshapen, a single nucleotide change leads to illness.

The desire to understand genetic diseases such as cystic fibrosis and sickle cell anemia was one motivation behind the Human Genome Project. This massive endeavor, completed in 2003, involved the identification of all the (approximately) three billion bases that make up human DNA. The project resulted in the identification of roughly 25,000 human genes; the function of many of these genes remains unknown. Now that the genome sequence is known, scientists are interpreting it and trying to determine the roles played by all the proteins. As this work progresses, they hope to figure out how the various proteins interact both with each other and with external factors and eventually to develop an understanding of how various genes contribute to human health.

Although we now know (and, as we shall see, cloning research played an important role in the discovery) that each cell contains essentially the same DNA, genome research has made it increasingly clear that only a subset of all possible proteins exists in a given cell at a given time. This implies that different sets of genes are turned on (or expressed) in different cells. For example, the genes expressed in nerve cells differ significantly from those turned on in liver cells, and the specific subset expressed determines how each cell behaves. Many scientists view the set of genes turned on in a particular cell, sometimes called the "gene expression program," as a key determinant of cell type. This suggests that changing the set of genes turned on in a particular cell may be one way to change its type: a possibility that has important implications for embryonic stem cell research.

Understanding the control of gene expression is an active area of research. In many cases, gene expression is regulated by proteins called "transcription factors." These proteins bind to specific DNA sequences (called promoters or enhancers) and either facilitate or block the expression of various genes. Modifications in the structure of chromosomes also play an important role in controlling gene expression. For example, small molecules known as methyl groups can attach to DNA and prevent transcription factors from binding at the appropriate locations, thus altering gene expression patterns. We will return to this topic later, when we examine the health problems of cloned animals, many of which are believed to result from faulty control of gene expression.

Cells at a glance

To understand the cloning process fully, we need not only to understand DNA and the way it codes for proteins but also the structure of the cells within which those proteins operate. Our discussion will focus on eukaryotic cells (the type of cell found in

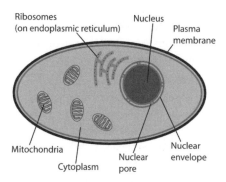

Figure 6 The major structures of eukaryotic cells, the type found in humans and other higher organisms. DNA is located in the nucleus.

almost all organisms other than bacteria and blue–green algae). These cells are small, invisible to the naked eye. The diameter of an average eukaryotic cell, such as a human skin cell, is 20 microns. (A micron is one millionth of a meter, so approximately 10,000 of these cells could fit on the head of a pin.) Some, such as blood cells, are round; others, such as neurons, are elongated but despite their differences, all eukaryotic cells have a number of similarities:

> They are surrounded by a semi-permeable membrane that separates the interior of the cell from its environment. This barrier, the plasma membrane, allows nutrients and some other substances to pass freely into and out of the cell, while blocking the passage of other materials. Protein complexes dot the membrane and control the transport of large molecules across it.

> They have a nucleus. This structure is typically found near the center of the cell and contains its DNA. The nucleus is surrounded by two membranes, each similar to the plasma membrane, which together form the nuclear envelope. This envelope is dotted with indentations, nuclear pores, which are the sites of transport into and out of the nucleus. Transport into the nucleus is mostly limited to transcription factors, proteins that regulate gene expression. Transport out of the nucleus is restricted, for the most part, to messenger RNA (mRNA), a precursor molecule that provides a template for protein synthesis within the cell.

> They contain ribosomes. These are the sites of protein synthesis within the cell and, interestingly enough, are not in the nucleus but in the cytoplasm, the name given to the cell's contents outside the nucleus. The ribosomes either float free or are attached to a structure called the endoplasmic reticulum. Once synthesized, a new protein leaves the ribosome to face one of many fates: it might function within the cytoplasm, it might be exported from the cell through a channel in the plasma membrane, or it might be imported into the nucleus, where it

could play a role in regulating gene expression and determining which proteins are synthesized in the future.

They contain mitochondria. These sausage-shaped structures are found throughout the cytoplasm and are often referred to as the "power plants" of the cell because they provide the energy the cell needs to function. This energy production is central to life itself but mitochondria are interesting to cloning scientists for another reason – they contain DNA.

Mitochondrial DNA complicates our picture of the cell and its genetic material. Instead of the nucleus containing a cell's full complement of DNA, it actually contains only a portion, while the remainder resides in numerous mitochondria sprinkled throughout the cytoplasm. This mitochondrial DNA (mtDNA) is quite specialized: it only contains genes that code for proteins that are used in the mitochondria. These proteins don't affect gene expression in the nucleus nor do they affect other cellular components. Scientists typically refer to nuclear DNA as "DNA" and explicitly mention "mtDNA" when referring to mitochondrial DNA.

The distinction between these two types of DNA is crucial for cloning. As we will see, cloned organisms share the same nuclear DNA but do not typically have identical mitochondrial DNA. Because mtDNA isn't thought to play a particularly large role in an organism's development or behavior (except, of course, ensuring its mitochondria function properly), the differences in mtDNA between cloned organisms are, typically, not considered important. This may well be the case but, as we will see, the distinction may have far-reaching implications, particularly in cases where cloning research crosses species boundaries. For example, consider a research project undertaken in China and published in 2003. A team of scientists, led by Hui Zhen Sheng at the Shanghai Second Medical University, transplanted human DNA into rabbit eggs from which the rabbit nuclear DNA had been removed.[2]

The resulting cell, which the scientists hoped to grow into embryonic stem cells, contained human nuclear DNA and rabbit mtDNA. Because this experiment blurred species boundaries by creating cells that were, in some critics' view, part-human and part-rabbit, it was heavily criticized on ethical grounds. The controversy this project generated suggests that the distinction between mitochondrial and nuclear DNA should be carefully considered when discussing cloning.

Cell division and the cell cycle

The population of cells that make up a single human being is in constant flux. At any given second, cells are dying and new cells replace them. The rate of turnover varies according to cell type. Some cells, such as neurons, are extremely long-lived, and may persist for years – even for an organism's entire life. Other cells

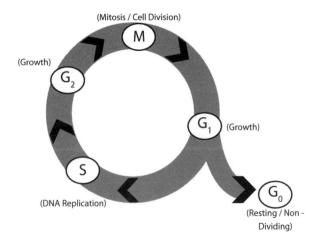

Figure 7 The cell cycle.

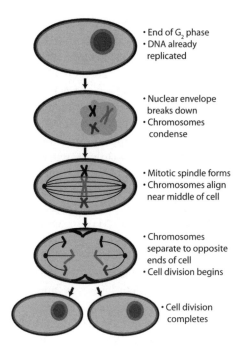

- End of G$_2$ phase
- DNA already replicated

- Nuclear envelope breaks down
- Chromosomes condense

- Mitotic spindle forms
- Chromosomes align near middle of cell

- Chromosomes separate to opposite ends of cell
- Cell division begins

- Cell division completes

Figure 8 Mitosis (the M phase of the cell cycle) is the process through which one cell divides into two identical daughter cells. To ensure each daughter cell is genetically identical, all the chromosomes line up and then segregate to opposite ends of the dividing cell.

have much shorter lives: red blood cells, which are responsible for the transport of oxygen and carbon dioxide through the body, live an average of 120 days, while the cells that line the human gut typically live only three to five days. New cells arise through a process of cell division in which a single parent cell divides into two daughter cells. This process, which is simple in concept but complex in execution, is part of a larger "cell cycle." Cells go through this cycle at different rates but every cell fits into one of its five major phases.

The cell cycle can be divided into two major components: mitosis and interphase. Mitosis is the portion of the cell cycle in which division actually occurs. It is short, lasting as little as one hour, but remarkably complex. The rest of the cell's time is spent in interphase. Cells that are not actively dividing may remain in interphase for days or years, waiting for a signal to start dividing.

Scientists have divided interphase into three primary phases. Two of these, G_1 and G_2, are primarily growth phases. In between comes the **S** phase; during this phase, the cell's nuclear DNA is replicated in preparation for division. G_1 accounts for most of the variability in cellular lifespan: some cells pause in G_1, entering a resting phase called G_0. Mitosis, or **M** phase, is the final stage of the cell cycle. It starts when the nuclear envelope, which normally separates the genetic material from the cytoplasm, disintegrates. Around this time, the chromosomes condense and, viewed through a microscope, become visible as individual structures. (During the rest of the cell cycle, chromosomes are thin and hard to see.) The chromosomes line up near the middle of the cell, attached to a structure called the mitotic spindle. Finally, the attachment between the paired chromosomes breaks and the two members of each chromosome pair separate to opposite sides of the cell. Finally, new nuclear envelopes form around the two sets of chromosomes and the cell divides. The result is two cells, each with identical genetic material.

This conservation of genetic material is a critical requirement for most cell division. But, it can't be used for all divisions. Sperm and eggs (germ cells) must be produced differently because they each contain only half of the chromosomes of a normal cell. Germ cells are created through a different cell division process: meiosis. In meiosis, the DNA is replicated once but two cell divisions occur. Thus, meiosis results in four cells, each with only one copy of each chromosome.

Mammalian development

Developmental biology is the study of how organisms grow and develop; how a single cell can give rise, in mere months, to a complex multicellular organism. Although many mysteries remain, developmental biologists can provide a fairly complete description of development in mammals and many other organisms. Cloning research has played an important role in the development of this understanding. As we examine the history of cloning research in the coming chapters, we will see that cloning was, at least in its early days, used by developmental biologists to answer specific scientific questions. Today, cloning research has moved beyond the boundaries of developmental biology, as scientists push toward new and controversial applications. Still, cloning remains, at its core, a variant of normal development and it is crucial to understand how mammals develop to understand how cloning works and how it could alter society in the future.

In humans, and in other mammals, development begins when an egg is fertilized by a sperm. Fertilization is not a simple, instantaneous, process. Rather, fertilization in humans involves three

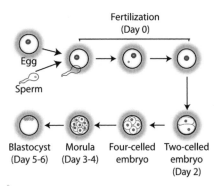

Figure 9 Human embryonic development from fertilization to the blastocyst stage.

distinct steps – penetration, activation, and fusion. Penetration occurs when a sperm navigates through the cells surrounding the egg and tunnels its way through the *zona pellucida*, a thick protective layer that surrounds the egg. Typically sperm penetration occurs in the oviduct, near the ovary. Development begins as the fertilized egg travels down the oviduct toward the uterus.

Successful penetration initiates of series of events collectively known as "egg activation." The first event is a change in the plasma membrane of the egg. This change, which occurs almost instantly following penetration, stops further sperm entering the egg, preventing a host of complications that would follow if a single egg was fertilized by multiple sperm. In mammals, penetration also triggers the completion of meiosis. Before sperm penetration, the egg contains two copies of each chromosome but following penetration the second meiotic division occurs, and half of the genetic material is eliminated. This leaves the egg with only a single copy of each chromosome and is a crucial step to prepare for a merging of the genetic material from the egg and sperm.

The final stage of mammalian fertilization is the fusion of the egg and sperm nuclei. This typically occurs about twelve hours after sperm penetration. During this time, the genetic material from the egg and the sperm remain separate, each in its own pronucleus. These structures slowly migrate toward each other as the egg prepares to complete fertilization and divide for the first time. During this migration, the 23 chromosomes from the egg and the 23 chromosomes from the sperm replicate, so when migration is complete, each pronucleus contains two copies of its own 23 chromosomes. When the pronuclei merge, two full sets of genetic material align on a single mitotic spindle and mitosis occurs, yielding two cells, each with a full copy of the new organism's genetic material. Once fertilization is complete, the new organism is called a zygote.

The next stage in mammalian development is referred to as cleavage. This is a process of cell division in which the zygote divides repeatedly and forms many distinct cells. Cleavage occurs without any appreciable growth, so each division simply divides the zygote into smaller and smaller component cells. The first division, as we have seen, occurs approximately twelve hours after sperm penetration and yields two identical cells. The second division also occurs symmetrically and yields four identical cells. If any of these cells separate at this point, identical twins will result. Typically all the cells remain together and ensuing divisions create embryos with eight cells, then sixteen and so on.

At the sixteen-cell stage, the embryo goes through a process called compaction, when the cells become much more tightly linked. The developing organism is now called a morula. The first evidence of cellular specialization is seen at this stage, when the embryo consists of sixteen cells. This specialization is based on location within the developing embryo. Cells on the outer edge of the morula divide asymmetrically toward the outside and go on to form the placenta and other extra-embryonic tissues. These cells are known as trophectoderm. Cells on the inside of the morula form the inner cell mass of the developing embryo. Eventually, cells from the inner cell mass will give rise to the cells of the mature organism.

Around day five or six in human development, the trophectoderm and the inner cell mass are distinct. The developing embryo is now called a blastocyst. At this stage, the embryo is nearing or has reached the uterus and consists of approximately 200 cells, only a small fraction of which make up the inner cell mass. After reaching the uterus, the embryo hatches out of its *zona pellucida*, which surrounded it during its trip down the oviduct. Assuming all goes well, the embryo, having reached the uterus and shed its *zona pellucida*, implants in the uterine wall around day eight or nine. The assumption that all goes well is not easily justified: although studying human development *in vivo* (in actual humans rather than test tubes or petri dishes) is challenging, estimates suggest that as many as 50 percent of fertilized human embryos fail to implant and are spontaneously aborted.[3] Researchers believe that this seemingly low implantation rate represents a quality control mechanism, preventing the implantation of unhealthy embryos that would not be likely to develop successfully.

Following implantation, development focuses on the placenta, the organ that connects the developing embryo with the mother and permits the transmission of nutrients during pregnancy. Placenta formation is crucial to successful development. It is also particularly relevant to cloning because, as we will see in later chapters, many of the health problems that occur in cloned animals are associated with improperly formed placentas.

The next major stage in the development of the embryo is "gastrulation." This occurs between days fourteen and sixteen, when the cells of the inner cell mass fold in on themselves to create distinct cell layers. These layers eventually form the three primary germ layers found in humans and other mammals: ectoderm, endoderm, and mesoderm. These layers have different developmental fates and their formation during gastrulation marks the first substantial differentiation of the cells from the inner cell mass. The differentiation that occurs at gastrulation is so crucial to life that Lewis Wolpert, one of the most distinguished developmental biologists of the last half-century has famously said, "it is not birth, marriage, or death but gastrulation, which is truly the most important time of your life."[4]

The other key event that occurs at the time of gastrulation is the appearance of the "primitive streak." The primitive streak is the first visible indicator of the differentiation that takes place during gastrulation and has often been invoked as a key milestone in the development of ethical guidelines for embryo research. The primitive streak, which appears around fourteen days, is pointed to as a precursor of neural development and thus a boundary to the earliest possible development of sentience by the embryo.

Gastrulation is quickly followed by neurulation, in the third week after fertilization, and organogenesis, in the fourth week after fertilization. By the end of the fourth week, the heart has formed and begun to beat, and arm and leg buds have appeared. At the end of the first month, a typical human embryo has increased in size nearly fifty times and is roughly five millimeters in length.

During the second month of development, in the process of "morphogenesis," the embryo begins to take on a more characteristic human shape. The arms, legs, and smaller extremities, including the fingers and toes, become visible. By the end of the second month, the embryo has increased in length to 2.5cm and weighs approximately one gram.

The third month includes the development of the major remaining organ systems. The start of the third month also marks the transition from embryo to fetus, a distinction made in the eighth week of development. Major milestones during the third month include development of the sensory organs and the framework of the nervous system. From this point, the organ systems are largely complete and the remaining months of development are dedicated primarily to growth.

The final six months of development are less relevant to cloning and will not be discussed in detail. The major emphasis is on growth of the developing fetus. By the end of the second trimester, the fetus typically weighs 600g and is over 30cm long. The third trimester includes even more growth, until the fetus is large enough to survive outside the womb. Neural development accelerates during the final trimester, as new neurons are formed and connections are made between neurons in the developing brain.

Why this matters

What does all this biology have to with cloning? Quite a bit. Cloning involves transferring the genetic material contained in the nucleus of a single cell into an unfertilized egg from which the genetic material has been removed and then activating this

PARTHENOGENESIS: NATURAL ASEXUAL REPRODUCTION

Although sexual reproduction is the dominant reproductive strategy among higher organisms, it is not the only one. Some species reproduce through an asexual technique: parthenogenesis. This means, quite literally, "virgin birth" and refers to a reproductive strategy in which females produce offspring without any male contribution.

Parthenogenesis occurs naturally in invertebrates such as water fleas and aphids, as well as some vertebrates, including lizards and salamanders. It can also be induced artificially in other species. In sea urchins, for instance, it can be triggered by placing unfertilized eggs in a solution of seawater and magnesium chloride.

In mammals, reproduction by parthenogenesis has not been reported (with the exception of animals genetically modified for the purpose of reproducing in this way). However, it is possible to activate an egg artificially leading it to start dividing as if it were fertilized. Such an egg may divide several times before development halts.

Although parthenogenesis is a rare reproductive technique, it is important to keep it in mind when discussing cloning, as they are both approaches to asexual reproduction. Indeed, scientists cloning animals must be careful to ensure the animals they produce are derived by cloning and not by parthenogenesis.

reconstructed embryo and coaxing it to develop into a new organism. A successful cloning procedure takes advantage of the knowledge of genetics and cellular biology developed over the last century and applies it to the creation of a new organism that must go through the development process. As we will see in the following chapters, the basics of biology we have examined in the preceding pages provide the parts list required to understand advanced cloning science and glean insights into how cloning may dramatically affect the future of human society.

Further reading

Most introductory biology textbooks will dedicate at least a chapter to each of topics introduced here. Almost any recent edition should be useful if you want a more thorough discussion of basic biology.

Several biographies mix the science introduced in this chapter with the stories of the lives of key scientists: Robin Marantz Henig tells the story of Mendel and the rediscovery of his work in the early twentieth century in *The Monk in the Garden: The Lost and Found Genius of Gregor Mendel, the Father of Genetics* (Mariner Books, 2001).

Many accounts of the discovery of DNA's structure exist. James Watson's first-person narrative, *The Double Helix: A Personal Account of the Discovery of the Structure of DNA* (Touchstone, 2001), provides a unique perspective and is worth reading. For those interested in another perspective, Brenda Maddox's *Rosalind Franklin: The Dark Lady of DNA* (Harper Perennial, 2003) may be of interest. Franklin, a contemporary of Watson and Crick, produced X-ray crystallography images that played an important role in the discovery of DNA's structure but did not receive much credit for her work.

An excellent introduction to the human genome project can be found in Matt Ridley's *Genome: The Autobiography of a Species in 23 Chapters* (Harper Perennial, 2000). Ridley dedicates one chapter to each of the twenty-three human chromosomes and tells a fascinating story that should appeal to anyone interested in how our genes work.

3
Dolly and her scientific predecessors

The announcement, in February 1997, of the birth of Dolly, the first mammal cloned from an adult cell, marked a clear dividing line in the history of cloning research. Before Dolly, cloning was a complex technique of limited use. It might have worked occasionally in lower organisms such as frogs; it might even have been possible to clone some mammals using cells taken from early embryos but most scientists viewed the cloning of mammals from adult cells as impossible. To paraphrase Lee Silver, Professor of Molecular Biology and Public Affairs at Princeton University, after Dolly, anything was possible.[1]

This chapter examines the history of cloning research, highlighting the steps and missteps taken *en route* to that day in February 1997 when suddenly, for developmental biologists, anything and everything was possible. The birth of a mammal cloned from an adult cell was a big deal because it meant that, for the first time, scientists could produce a genetic copy of another living animal. The donor could be a known quantity, chosen for its specific qualities or desirable attributes. Cloning from embryonic cells was less exciting because you were, by definition, creating a genetic duplicate of an embryo whose abilities and attributes were unknown.

As is the case with so much of science, the story of Dolly's cloning is not a simple progression from experiment to experiment. Rather it is a story of a few stunning, and often unpredictable, scientific breakthroughs separated by long periods of frustration.

These scientific advances were intertwined with public concern and controversy over the proper scope of biomedical research and shaped by allegations of fraud, both in the public sphere and within the close-knit scientific community. This story will set the stage for understanding not just the cloning of Dolly but also current cloning research.

Developmental biology preludes to cloning

When scientists talk about cloning today, they are referring to a specific technique. This technique, known as "somatic cell nuclear transfer" (SCNT), involves the transfer of genetic material into an egg from which the genetic material has been removed. This technique was first proposed in 1938, when it was described by Hans Spemann, the first embryologist to win a Nobel Prize, as an "experiment which appears, at first sight, somewhat fantastical."[2] Our story starts not with Spemann's proposal but some fifty years earlier with August Weismann, whose theory inspired Spemann and countless others.

Weismann, a zoologist at the University of Freiburg, sought to explain why cell differentiation seemed to go in only one direction. He knew that a single fertilized egg could give rise to every cell in an organism but never observed a differentiated cell morphing from one type to another. To explain these observations, Weismann proposed that during cell differentiation the genetic material in each cell was reduced in quantity. Thus, once differentiation was complete, each cell would contain only the genetic material specific to its particular type: brain cells would contain only the information needed to be brain cells and not that needed to form blood or liver cells. The same would be true for each of the cell types in an organism. This, the "germ-plasm theory of heredity," proposed that reduction in genetic material characterized essentially all cell division, so that once a fertilized egg divided to form a two-cell embryo, each cell should contain only enough genetic material to create half an organism.

Weismann's theory and the experiments it inspired provide an excellent picture of the scientific method at work. The theory immediately suggested a number of experiments that would either support or refute it. This is crucial. To be useful, scientific theories must not just explain existing observations but also make testable predictions and be logically falsifiable. (That is, there must be some potential experiment or observation that could prove a theory incorrect.) These characteristics are three of the key factors that distinguish scientific theories, such as Newton's theory of gravity, Einstein's theory of relativity and Darwin's theory of evolution, from pseudoscience, such as astrology or intelligent design. A scientific theory need not be correct to prove useful. Weismann's theory, as we will see, was wrong but by generating testable predictions, it inspired numerous scientists and advanced embryological research.

The first experiment that directly tested Weismann's theory was performed by the German embryologist Wilhelm Roux. Roux studied frogs – a favorite organism for embryologists due to their large eggs – and decided to see what happened when one half of a two-cell embryo was destroyed. If Weismann's theory was correct, the remaining cell should grow into half a frog. If Weismann was wrong and each cell contained a full set of genetic material, the remaining unharmed cell might develop into a complete frog. In the spring of 1887, Roux collected fertilized frog eggs from a nearby pond and brought them to the lab. He waited until they had divided once, creating two-cell embryos. Roux then destroyed half of each embryo, by puncturing one cell with a hot needle. The results were just as Weismann had predicted. The remaining cell continued to develop but never became a complete frog. Development halted, leaving fragments that looked just as one might imagine half a frog embryo to look.

Roux's apparent success inspired additional research, and this work clouded the picture. Hans Adolf Eduard Dreisch, working several years later, repeated Roux's experiments using sea urchins. Because sea urchin eggs were smaller and harder to manipulate than frog eggs, Dreisch used a different method. Rather than destroying one cell with a hot needle, he separated the two cells by shaking the embryo vigorously in salt water. To Dreisch's surprise, his results contradicted both Weismann's theory and Roux's results. Instead of developing into sea urchin halves, each cell gave rise to a complete and healthy, although somewhat small, sea urchin. The same result was seen when Dreisch started with four-cell embryos. Rather than sea urchin quarters, Dreisch ended up with four complete, but small, sea urchins. These results irrefutably contradicted Roux's earlier work and Dreisch searched for an explanation. He believed that Roux's use of a hot needle might explain the differences, as perhaps the needle had somehow damaged the remaining cell and prevented it from developing normally. He was unable, however, to separate two-cell frog embryos by shaking to test this hypothesis.

The resolution came in an experiment by Spemann, who in 1902 managed to separate salamander embryos at the two-cell stage. Rather than shaking the cells, Spemann sliced them apart. To accomplish this delicate task, he created a noose using hair from a newborn baby and gently tightened it until the two cells were completely separated. Spemann found that the two cells developed into complete organisms. These results supported Dreisch's work with sea urchins and refuted Weismann's theory.

Although Weismann's germ-plasm theory was no longer tenable, Roux's results with frog embryos remained a contradiction until 1910. The mystery was finally solved when Jesse Francis McClendon, at Cornell University, separated a two-cell frog embryo and observed successful development of complete, rather than partial frogs. In retrospect, Roux's mistake is obvious. According to Robert McKinnell, a biologist who wrote about the history of cloning technology in the late 1970s, Roux conducted a good experiment but misinterpreted his results.[3] The abnormal development of the remaining unaffected cell was due not to any reduction in its genetic potential but to the impact of being attached to a dead cell. The dead cell that Roux had punctured with a hot needle formed a roadblock and prevented the remaining cell from developing to its full potential.

Spemann continued his research and went on to show that cells from a variety of early embryos could give rise to entire organisms. Through the creative use of his baby-hair nooses he was able to separate cells up to the sixteen-cell stage. This work showed that the genetic material found in individual cells retained the potential to direct the development of an entire organism. The sixteen-cell stage, however, is still very early in development. It was Spemann's desire to repeat these experiments with more developed cells that led to his suggestion, in 1938, of the nuclear transfer procedure.

The first cloned frog

Although Spemann wrote that he could see no way to perform the fantastical experiment he proposed, research progressed rapidly. In 1952, fourteen years after Spemann's suggestion and just a year before the structure of DNA was discovered, researchers in Philadelphia used nuclear transfer to clone frogs. The scientists, Robert Briggs and Thomas King, worked at the Institute for Cancer Research, now known as the Fox Chase Cancer Center, and, although they were unaware of Spemann's proposal, they used a method nearly identical to the one he had outlined.

Briggs and King worked with the northern leopard frog, which is common in ponds throughout North America. Much like their predecessors in the early twentieth century, they designed their experiments in the hope of understanding cellular differentiation.

In particular, they hoped to use nuclear transfer to test the differentiation status of various cell types and determine whether or not the nuclei in these cells were irreversibly differentiated. To begin their research, Briggs and King worked with cells from the blastula stage, roughly equivalent to the blastocyst stage in mammalian development. In frogs, cells at this stage are essentially undifferentiated and Briggs and King were hopeful that these cells would, like undifferentiated cells from earlier stages, give rise to complete organisms.

Despite this hope, the project faced significant challenges. The first was funding. External funding agencies, such as the U.S. National Institutes of Health or the Medical Research Council in the United Kingdom, typically prefer to fund research with a good chance of success. The frog cloning project, however, was novel and seemed far-fetched. Briggs' first grant proposal was rejected with one reviewer characterizing the project as a "harebrained scheme with little chance of success."[4] Despite this initial setback, funding was eventually obtained and the work began.

The next challenge was technical. It was not clear that the nuclear transfer procedure was possible. Even if it was, the process seemed likely to damage the delicate biological materials involved and perhaps render the cloned embryos unable to develop. The first step was to remove the genetic material from an unfertilized egg. To accomplish this, the scientists first pricked the egg with a clean glass needle. This prick, much like penetration of the egg by a sperm, set in motion the process of egg activation. As part of this process, the chromosomes in the frog egg move toward the surface of the cell. This movement allowed King, who did most of the micromanipulation, to capture the egg's nucleus in a hollowed-out needle and remove it from the cell. (An egg whose genetic material has been removed in this way is said to be "enucleated.") Next, the jelly coat surrounding the egg was removed. After these manipulations, the enucleated egg was ready for transplant.

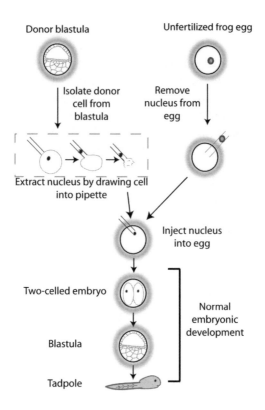

Figure 10 The method used by Briggs and King to clone the first frog. The extraction of the blastula cell nucleus (dashed box) is magnified more than the rest of the diagram.

Isolating the donor nucleus posed an even bigger challenge. Frog blastula cells are small and difficult to work with and no established technique existed to remove a cell's nucleus without damaging it. To isolate the nucleus, Briggs and King developed a technique using a customized micropipette (a specialized tool similar to an eye dropper that scientists use to transfer tiny quantities of liquid from one place to another). The openings on Briggs and King's special micropipettes had diameters larger than a typical frog nucleus but somewhat smaller than the diameter of the blastula cell. To isolate the nucleus, the entire cell was slowly drawn up into the micropipette; because the opening was smaller than the diameter of the cell, the cell was compressed and distorted. Eventually, the cell membrane ruptured and the smaller nucleus remained intact in the tip of the micropipette. It could then be injected into the previously enucleated egg.

Although perfecting the technique was challenging, Briggs and King eventually found success. In their landmark 1952 publication describing the cloning process, they reported transferring blastula stage nuclei into 197 enucleated eggs.[5] Many of these reconstructed eggs started to divide and develop and 32 percent successfully reached the blastula stage – the stage from which the initial donor cells had been taken. Most of these embryos continued to develop and many eventually grew into normal healthy tadpoles. The appearance of these tadpoles was taken as proof that Spemann's fantastical experiment could work. Cloning had come of age as a biological research tool.

More frogs and more confusion

Briggs, King, and others quickly followed up this exciting breakthrough with additional studies. Scientists determined that other amphibians could also be cloned from embryonic cells and explored cloning using more advanced cells. King recalls rampant optimism following the first cloning success, but it quickly became clear that cloning efficiency declined rapidly when donor nuclei from more mature cells were used.[6] This drop-off was consistent across a variety of different amphibians. Both the northern leopard frog and the African clawed frog could be cloned at relatively high efficiency using cells from one-day-old embryos but neither could be cloned efficiently when cells were taken from embryos much older than two days. These results suggested that Weismann might have been partially correct. It seemed the developmental potential of cells changed as they matured. Early embryonic cells remained flexible and could direct the development of a new organism but mature differentiated cells were restricted in their potential and could no longer generate an entire organism.

This understanding was challenged most significantly by an experiment performed by John Gurdon at Oxford University. Gurdon worked with the African clawed frog, and had previously reported its successful cloning using the nuclear transfer procedure developed by Briggs and King. More interesting was Gurdon's report that he successfully cloned frogs using donor cells taken from the intestinal lining of tadpoles. The tadpole stage is much further along in development than the blastula and, presumably, most of the cells in the tadpole's intestine are differentiated. Although this procedure was only rarely successful, it seemed to prove that cloning using differentiated cells was possible. In an influential 1962 report, Gurdon showed that some frogs cloned from tadpole intestinal cells grew to sexual maturity and were themselves fertile.[7]

To clone frogs from these more differentiated cells, Gurdon used a more complex nuclear transfer procedure, serial transfer. He first transferred a nucleus from an intestinal cell to an enucleated egg. After this cloned embryo had begun to develop normally, he transferred a nucleus from a cell in the developing embryo to a second enucleated egg. This second cloning step, for reasons that were not fully understood, seemed to promote healthier development when using donor nuclei from more differentiated cells.

Gurdon's conclusions proved controversial and difficult to replicate. Some scientists thought these clones might have originated through transfer of nuclei from primordial sperm or egg cells rather than differentiated cells. During development, these primordial germ cells migrate down the intestinal lining of the frog species Gurdon studied, making up a small percentage of the cells found there. The low success rate Gurdon reported suggested to some that his success came from the rare cases where his donor cell was a primordial germ cell rather than the differentiated cells more typical of the intestinal lining. If this was the case, Gurdon's frogs were cloned not from fully differentiated cells but from much less mature cells. As Marie Di Berardino, a scientist who worked in Briggs' lab when frogs were first cloned has written, interpretation of these experiments was difficult because no unequivocal marker of cell differentiation was used when selecting the donor cells.[8] In later experiments, many scientists found that indisputably differentiated cells could provide donor nuclei that directed partial but not complete development. In particular, tadpoles occasionally resulted when skin or blood cells provided the donor nuclei. These tadpoles, however, invariably died without maturing into adult frogs.

The result of this mix of experiments was confusion, rather than clarity. Scientists could say conclusively that donor nuclei from embryonic cells could successfully direct development and yield healthy adult frogs. They also knew that donor nuclei from differentiated cells could give rise to tadpoles but were not certain if these differentiated cells could give rise to adult frogs. In no clear case had nuclear transfer of an adult cell given rise directly to a healthy fertile adult.

Mammalian cloning: fact or fiction?

Adapting the cloning technique pioneered by Briggs and King for use with mammals posed many challenges. First and foremost, the mammalian egg is much smaller than a typical amphibian egg. While a frog egg is one to two millimeters in diameter, a mouse egg is less than one hundred microns. As nuclear transfer in frogs is a difficult technique requiring specialized tools, it is not surprising that duplicating the process with an egg a thousand times smaller in volume posed significant challenges.

The field did advance, however. Cells from eight-cell rabbit embryos were isolated in 1968, showing that Spemann's early work with amphibians could be replicated in mammals. Derek Bromhall, working at Oxford University, transferred nuclei from embryonic rabbit cells into enucleated rabbit eggs and reported observing some embryos that developed to the blastocyst stage. Bromhall did not transfer these embryos into the wombs of surrogate rabbits, and thus it is not clear if they would have developed normally. Despite these advances, the adaptation of the nuclear transfer technique to work with mammals was proceeding quite slowly.

Given this slow progress, scientists reacted with great shock when an apparently huge advance was reported not in the peer-reviewed scientific literature but in a popular (and seemingly non-fiction) book. This work, *In His Image: The Cloning of a Man*, was written by David Rorvik, a respected science journalist, and published in 1978 by a reputable publishing house, J.B. Lippincott. In the book, Rorvik told the incredible story of an eccentric millionaire, referred to as Max, who had apparently contacted him and asked for his assistance.[9] The author's task was to help Max create a male heir by cloning himself. In the story, Rorvik helped Max recruit a scientist – code-named Darwin – to work with him. In a short period of time, working at a hospital Max owned on a remote Pacific island, Darwin and another scientist reportedly overcame numerous obstacles that had stymied others working in the field. In the book, Rorvik reports how Darwin successfully transferred nuclei from Max's adult "body-cells" into eggs unknowingly taken from women visiting Max's hospital. In the story, Darwin and his colleague leapfrog years of fertility research and successfully grow these cloned embryos *in vitro* and transfer them into the uterus of a surrogate mother – a young virgin named

Sparrow. According to Rorvik, Max's clone was born in a small hospital in the United States in December 1976.

At the point when Max's clone was supposedly born, no child had ever been born through *in vitro* fertilization nor had normal development ever been seen following nuclear transfer using embryonic, much less adult mammalian, cells. Even in amphibians, in which cloning had been studied for some 25 years, there was not a single undisputed case of a healthy adult resulting from simple nuclear transfer of an adult cell. In short, from a scientific perspective, the story Rorvik told was nothing short of outlandish.

Most scientists were highly skeptical and loudly proclaimed the work a hoax, pointing out numerous scientific errors in the text. Rorvik, however, while acknowledging the story was shocking and seemed unlikely, asserted it was true. The general public wasn't sure who to believe. At first glance the book appeared scientifically thorough. It contained nearly thirteen pages of references, many of which included detailed descriptions of advances reported in the peer-reviewed literature. The book generated intense publicity and found a willing audience; climbing the non-fiction bestseller list in both the United States and the United Kingdom. In the United States it spent six weeks on the *New York Times* non-fiction best-seller list in May and June 1978, reaching as high as ten at one point.

In the end, one of the book's many references proved its downfall. Derek Bromhall, the Oxford biologist mentioned briefly above, sued Rorvik and J.B. Lippincott, claiming the book defamed him. Bromhall argued that, by referencing his work without permission, Rorvik suggested that his research aimed at cloning humans. Eventually a judge declared the book a hoax and Lippincott and Bromhall settled out of court. A key piece of evidence was a letter sent from Rorvik to Bromhall asking for details on his research, dated some five months after Max's clone had ostensibly been born. Lippincott publicly apologized and admitted that it now believed the book was untrue. Sources reported that Bromhall received approximately $100,000 from the publisher.

Although this settlement discredited the book, it came four years after publication. By then, its influence had already been felt. Some 95,000 copies had been sold in hardcover. Rorvik made nearly $400,000 on the book and the publisher earned more than $700,000. Many of the general public were left with the impression that even if humans had not been cloned in this case, it might not be far off.

Scientists worked to dispel this notion, claiming that human cloning was unlikely, if not impossible. They testified in congressional hearings, decrying Rorvik's work as fiction. But the public was not in a trusting mood. Recent excitement and fear over scientists' growing ability to genetically manipulate organisms and the quickening pace of biomedical research in general led some to question these assurances.

The mouse cloning controversy

These claims that mammalian cloning in general, and human cloning in particular, were far in the future were soon challenged once again. This time the challenge came not from an out-of-the-blue book but in a format more familiar to scientists, an article in a prestigious peer-reviewed journal. In January 1981 – while the Bromhall/Rorvik lawsuit was working its way through the courts – Karl Illmensee, a star developmental biologist working at the University of Geneva, and Peter Hoppe at the Jackson Laboratory in Bar Harbor, Maine, published a scintillating report of mouse cloning. This report came out in *Cell*, one of biology's most prestigious journals and, despite their surprise, scientists had no reason to doubt the results. Before an article is published in *Cell*, or a similar journal, the journal's editor sends it out for review to several experts in the field. These reviewers examine the article and report back to the editor recommending whether or not the paper should be published and suggesting alterations and improvements. In a prestigious journal, like *Cell*, only a small fraction of manuscripts (typically less than 10 percent) are eventually selected for publication and these papers are thought to be among the most important and rigorously conducted in the field.

The report was a shock. Illmensee and Hoppe described successfully repeating Briggs and King's initial frog cloning experiment using mice – the standard organism of mammalian developmental biology.[10] In their published report, Illmensee and Hoppe described experiments transferring nuclei from blastocyst-stage mouse embryos into fertilized enucleated mouse eggs. They used two types of donor cells: cells from the inner cell mass that eventually go on to form the organism and cells from the trophectoderm, which form the placenta. Nuclei from the trophectoderm led only to abnormal development. On the other hand, nuclei taken from cells in the inner cell mass led, in some cases, to normal development. According to their report, of the 363 eggs that were injected with these nuclei, 142 survived the micromanipulation procedure. Ninety-six of these began to divide and reached the two-cell stage. Half of these embryos successfully reached the morula and blastocyst stages. Sixteen of these apparently normal embryos were transferred to the uteri of female mice and three healthy cloned mice were born. Two of these cloned mice were mated with non-cloned mice and with each other and shown to be fertile.

Scientists, although surprised, celebrated the results as a major breakthrough. Illmensee had a reputation as a talented technician, and it seemed plausible that his skill had allowed him to succeed where others had failed. In short, his name and growing reputation lent instant credibility to the report. This success was short-lived, however: the report was questioned and essentially discredited, damaging both Illmensee's and Hoppe's credibility and hindering their careers.

Replication is a key element of the scientific method. Although not all experiments are repeated by other scientists, many of the most important results are. This replication verifies the validity of previously reported experiments and creates a starting point for follow-up research. In the case of Illmensee and Hoppe's cloned mice, this replication was problematic. Numerous scientists tried to repeat Illmensee's experiments but failed.[11] Furthermore, Illmensee refused to demonstrate the nuclear transfer technique to others, even other members of his own laboratory.

At first, scientists who couldn't repeat the cloning experiments tended to blame themselves. Illmensee's skill with nuclear manipulation was well-known and suspected to be crucial to the experiment's success. Repeated failed attempts at replication, combined with Illmensee's unwillingness to help raised suspicions. Eventually scientists in Illmensee's own lab began to suspect something was amiss. Finally, in January 1983, two years after the cloning paper was published, members of Illmensee's lab challenged his results, essentially charging him with fraud.[12]

These charges touched off numerous investigations. The most thorough, run by Illmensee's employer, the University of Geneva, was undertaken by a commission of experts in the field. They focused not on the cloning experiment but on a related study. This commission found "no compelling evidence of falsification of data" and thus found Illmensee innocent of fraud. The report, however, was far from complimentary of his experimental procedures and his documentation. Indeed, they found that his records contained "numerous corrections, errors and discrepancies." The commission concluded that the experiments they examined were "scientifically worthless" and suggested that Illmensee repeat them with outside collaborators present.[13] Although Illmensee was reinstated to his position following this report, the faculty of the University of Geneva refused to accept the commission's conclusions. Illmensee eventually left the University in 1985, amidst reports indicating that his fellow professors had recommended his contract not be extended. Although officially cleared by the commission, the charges of fraud were never fully resolved.

Resolution became less important as the excitement over cloning faded and scientists looked in other directions. Two scientists, James McGrath and Davor Solter, who had been among those attempting to replicate Illmensee and Hoppe's experiment, played a prominent role in this shift. Working at the Wistar Institute in Philadelphia, they found that they could routinely transfer nuclei from one-cell mouse embryos to other one-cell mouse embryos. This procedure seemed hardly to affect development but transferring nuclei from more advanced cells proved difficult. They did report some success using donor nuclei from two-cell embryos, where they observed development to the blastocyst stage but no further. When more advanced cells were used as nuclear donors, successful development to the blastocyst stage was not seen. McGrath and Solter transferred nuclei into enucleated fertilized eggs (as Illmensee had reported) rather then unfertilized eggs. At the time, the significance of this methodological choice was not grasped but in retrospect, it proves crucial. In any case, they concluded their 1984 report of this research with a strongly worded statement that would shape the field. They wrote that their and other recent results "suggest that the cloning of mammals by simple nuclear transfer is biologically impossible."[14]

Although, in retrospect, McGrath and Solter overstated their case, their conclusion represented the mainstream view at the time. Developmental biologists and the agencies that funded them, by and large, had lost interest in cloning. This shift was due in part to the uncertainty surrounding the famous mouse cloning paper. It was also due to the wide range of other questions available for scientists to address. Genetic engineering was rapidly advancing and scientists were on the road to creating mice tailor-made to answer specific scientific questions. Scientists had also discovered that, before reproduction, maternal and paternal DNA was modified in a gender-specific manner and that successful development required contributions from both parents. This phenomenon became known as imprinting. Today, imprinting has important implications for cloning researchers and may explain why some cloned animals are abnormal. In the 1980s it was an exciting new field – with verifiable results – that attracted developmental biologists who may otherwise have studied cloning. In contrast, cloning was a field with no clear future. It had been three decades since Briggs and King cloned the first frog, yet not a single mammal had been verifiably cloned.

Cloning on the farm

Solter and McGrath were wrong. Mammals can be cloned. And what's more, mammals can be cloned not just from embryonic cells but also from clearly differentiated adult cells. This dramatic turnaround began in the mid 1980s and culminated in the late 1990s, when Dolly's birth was announced.

Although the questions addressed were similar, this research was carried out by an almost entirely distinct research community. As we have seen, following the controversy over mouse cloning, mainstream developmental biologists lost interest in the field. This was not the case for animal scientists, a group of researchers largely based at agricultural research institutes and interested in the production of livestock, particularly cattle. Unlike previous cloning scientists, who had been interested in the technique primarily as a window to understand development, these researchers had more practical aims. Cattle are big business. Both milk production and meat quality vary significantly from cow to cow. Not surprisingly, cows that produce large quantities of milk or have high quality meat are more valuable. For these reasons, cloning high-quality cows seemed to offer an attractive business proposition.

Thus, commercially driven research proved Solter and McGrath wrong. Two research groups played key roles. The first, actually more of an individual than a group effort, was led by Steen

Willadsen, a Dane working at a British agricultural research center near Cambridge. Although Willadsen's eventual goal was to work with cows, he did his initial research with sheep, as they are cheaper and easier to study. Ironically, Willadsen's first successful cloning experiment, and the first undisputed cloning of a mammal by nuclear transfer, took place in March 1984, nearly nine months before Solter and McGrath's claim that mammals could not be cloned was published.

Willadsen's work essentially duplicated Briggs and King's original protocol but working with the diminutive sheep eggs required a number of modifications. Crucially, he chose to work with unfertilized eggs instead of the fertilized eggs used by Solter and McGrath in their unsuccessful attempts to clone mice. This proved a fortuitous choice. Rather than injecting the donor nucleus directly into the egg, which he feared might harm the delicate eggs, Willadsen placed an entire donor cell just outside an enucleated egg and fused the two cells. This fusion, which was brought about by a small jolt of electricity, joined the two cells together and brought the donor nucleus into the enucleated egg. Although the donor nucleus brought some cytoplasm and other cellular components with it, the egg is so much larger that its cytoplasm dominates the reconstructed egg.

The growth of the cloned sheep embryos presented the next challenge. To accomplish this, Willadsen took advantage of a technique he had developed previously. He wrapped the nascent embryos in agar – a jelly-like substance – and transferred them into the oviducts of female sheep. Embryos typically undergo their early development in the oviducts and Willadsen essentially used these sheep as short-term surrogates to duplicate this developmental environment. He tied the oviducts of these ewes shut, so the embryos could not reach the uterus, and retrieved them approximately five days later. After retrieving the embryos, Willadsen removed the protective agar layer. If development appeared to be proceeding normally, he transferred the embryo into the uterus of a second surrogate. The very first time Willadsen undertook this experiment, three of the cloned embryos developed to term.[15] Although he had used nuclei from undifferentiated cells taken from eight-cell embryos, the research showed that mammals could indeed be cloned by nuclear transfer.

The other group trying to clone large domestic animals was led by Neal First and based in Madison, Wisconsin. These researchers, who worked at the University of Wisconsin's College of Agricultural and Life Sciences, worked directly with cows and, using techniques similar to those used by Willadsen, reported successful cloning in 1987. Like Willadsen, these researchers used electric fusion to join enucleated eggs with donor cells. They explored several maturation strategies but found the most success when the cloned embryos were protected in agar and allowed to develop in the oviducts. After four or five days, the embryos were retrieved and transferred to the uteri of dairy cows. Success rates were low in general. First and his colleagues attempted the fusion procedure 558 times and ended up with twenty-three embryos that developed to either the morula or blastocyst stage.[16] They transferred nineteen of these embryos to surrogates and ended up with two live cloned cows.

Roslin and the quest for transgenic cows

Both Willadsen's work cloning sheep and First's work cloning cows set the stage for the cloning of Dolly. However, nearly another decade passed before Dolly was born in 1996. As in other livestock cloning efforts, the Roslin Institute, where Dolly was born, was outside the scientific mainstream. Although today it is well-known, before Dolly's birth few researchers had heard of it and even fewer were aware of the research underway there.

Like the cow cloning research in Wisconsin, the cloning research at Roslin was driven by commercial aspirations. This time, however, the commercial interest came not from the possibility of cloning prize cows but from the pharmaceutical industry. The eventual goal – and still a key driver of cloning research today – was the creation of cloned cows that produced valuable therapeutic compounds in their milk. During the 1980s, scientists learned how to insert specific genes into cells grown outside the body in cell culture and even, in some cases, directly into embryos. Their understanding of which genes were turned on in which cells also grew immensely. Together, these developments allowed them to insert genes coding for therapeutically useful proteins, such as insulin or a blood clotting factor, into cells along with control sequences that ensured these proteins would be produced in an animal's milk. Organisms, such as these, that contain DNA from some other species are called transgenic. Although Dolly herself was not transgenic, the goal behind the research that led to Dolly's cloning was the creation of transgenic sheep.

Such a goal need not necessarily lead to cloning and, indeed, Ian Wilmut, one of the scientists who led the effort to clone Dolly, worked for many years on alternative techniques to produce transgenic sheep. Before the research leading to Dolly, existing techniques to create transgenic farm animals were rather inefficient. DNA fragments were directly injected into cells, where they were randomly incorporated into an organism's nuclear DNA in a small number of cases. Often, the new DNA wouldn't be incorporated at all. Sometimes it might incorporate in the middle of another important gene, blocking its function. Because of these and other complications, the process was successful in perhaps one out of one hundred cases. To make matters worse, it typically wasn't clear if the animal expressed the transgene properly until after its birth. This meant that surrogate mothers had to be found for hundreds, if not thousands, of embryos. For mice, this wasn't a big deal: for cows, it was an expensive proposition.

One strategy to create transgenic cows more efficiently would be growing a healthy adult organism from a single genetically modified cell. Scientists had accomplished this trick in mice, using specialized undifferentiated cells, called embryonic stem cells. We'll talk more about these cells later, as their human equivalents are potentially the key to the nascent field of regenerative medicine. For now, it suffices to say that these cells had not been isolated from cows or other large domestic animals. Theoretically, these valuable proteins could be produced in mice's milk through embryonic stem cell technology but, for obvious reasons, this is not an attractive commercial proposition.

The big advantage of embryonic stem cells over fertilized eggs was that they could be grown and maintained in cell culture. This simply means that they can grow outside the body, typically in a small plastic tray, called a petri dish, when provided with appropriate nutrients. If Wilmut could insert his transgene into cultured cells, rather than fertilized eggs, it would be a huge benefit. In culture, it is relatively easy to check and see if the transgene has been successfully integrated into the cell. This meant that Wilmut would be working only with cells that contained the desired gene. Another advantage of embryonic stem cells – and this property is so far thought to be unique to them – is that after growth in culture, they can be used, in mice at least, to direct the development of a new healthy organism. Without embryonic stem cells, Wilmut seemingly had no path forward. He could grow and modify other more differentiated cells, such as skin or mammary gland cells, in culture but he had no way to use these cells to create a new organism.

Cloning offered an alternative strategy. Although no mammal had ever been cloned from a differentiated cell, Wilmut saw that such cloning, if possible, might provide a solution to his problems. If he could clone from a cell grown in culture, it should theoretically be possible to modify the differentiated cells in culture and then clone only from a population of cells expressing the transgene. If all went as planned, these cloned organisms should produce the desired protein in their milk. Many would have dismissed this protocol out of hand. After all, scientists had been studying nuclear transfer in amphibians since the 1950s and in mammals since at least the 1970s and not a single report of a healthy adult organism produced by cloning from an adult cell had been confirmed. Wilmut did have one reason to hope. At a scientific meeting in 1986, he had heard a rumor that Willadsen, the scientist who first cloned sheep from embryonic cells, had also successfully cloned cows using cells from much older embryos.[17] This work, which was never published, suggested that older differentiated and perhaps cultured cells could provide donor nuclei for successful cloning experiments.

Megan, Morag, and Dolly

With this slight glimmer of hope, Wilmut convinced his sponsors the work was worth funding. He also made a decision that would prove crucial to the project's success. Based on preliminary research suggesting the cell cycle might be an important factor in determining the success of cloning experiments, Wilmut decided to hire a cell cycle expert. Keith Campbell, then a postdoctoral researcher at the University of Dundee in Scotland, saw Wilmut's advertisement and moved to Roslin to work on the cloning project.

As described in Chapter 2, all cells are in one of several phases of the cell cycle. These phases include G_1 and G_2, where the cell is primarily growing. They also include the S phase, when a cell's DNA is duplicated in preparation for division, which occurs in the M phase. There is also a resting stage, G_0, where some cells remain for long durations. Campbell's task was to determine which of these cell cycle phases was appropriate for donor nuclei to be used in cloning experiments.

To accomplish this task, Campbell completed a series of nuclear transfer experiments using cow embryonic donor cells taken from cells in both G_1 and G_2.[18] Campbell fused these donor cells with two types of enucleated eggs: some activated simultaneously with the nuclear transfer, as would be the case during fertilization, and some activated ten hours prior to nuclear transfer. This complex experiment aimed to tease out the effect of a key protein on development following nuclear transfer. Among the findings of this study, one stood out for its implication on future cloning research: only diploid cells (those in G_1 or G_0) could serve as successful donor cells when the enucleated egg was activated simultaneously with transfer.

This work was carried out with cow cells and the cloned embryos were only allowed to develop to the blastocyst stage, but the results encouraged Wilmut and Campbell to continue the work in sheep. As cells growing in culture can easily be induced to enter G_0, by reducing the nutrients available for growth, Wilmut and Campbell decided to use G_0 cells as their diploid cell population. In their experiment, they used three types of donor cells.[19] One of these sources was early embryonic cells, like those Willadsen had used in his successful sheep cloning a decade earlier. However, Wilmut and Campbell also used cells extracted from early embryos and grown in culture. Some of these cells were taken just after they had been removed from the embryo, while they still appeared undifferentiated, and some were taken from cells that had grown in culture for a significant period of time and had clearly differentiated. All the cells in culture were induced to enter G_0 by nutrient starvation.

The nuclear transfer was accomplished using the technique Willadsen had previously developed. The donor cells were fused with enucleated unfertilized eggs through the use of an electric current and early embryonic development took place in the oviducts of surrogate sheep. Embryos that developed successfully were transferred again to new surrogates where the embryos were left to develop, hopefully, to term.

Given previous difficulties cloning with even partly differentiated cells, it seemed likely that only the early embryonic cells would prove to be functional donors. To Wilmut and Campbell's surprise, this was not the case. In fact, the source of the donor cell seemed almost irrelevant. Perhaps the use of cells in G_0 was the trick. In particular, when the donor cells were induced to enter this resting state before transfer, clearly differentiated cells that had grown and replicated up to thirteen times in culture could yield cloned embryos that developed successfully to the blastocyst stage. Cloning works because components of the egg cytoplasm reprogram the DNA taken from the donor cell, allowing it to guide development. These results led Campbell to hypothesize that the still poorly understood process of reprogramming might be easiest with cells in the G_0 phase, thinking the resting status of these cells might predispose them to the process. In total, thirty-four

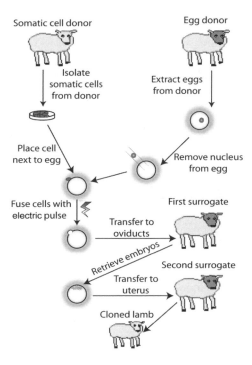

Figure 11 A rough outline of the process used to clone Dolly, Megan, and Morag.

embryos developed normally and were transferred to recipient ewes. From these, five live births resulted. Two lambs died shortly after birth and one died at ten days. The remaining two, Megan and Morag, survived. As it turned out, the two surviving lambs were identical twins, having been cloned from the same embryo.

Megan and Morag were the first mammals cloned from differentiated cells. Although their births didn't generate the excitement that Dolly's would a short time later, from a practical point of view, it was more important and surprising. First, it proved that differentiated cells could provide donor nuclei for nuclear transfer. This finding refuted a belief that had been gaining traction since the 1950s when Briggs and King had realized how quickly cloning efficiency declined as donor cells aged. Now suddenly, cells that had been grown and duplicated in culture up to thirteen times over several days could be used in cloning experiments. Furthermore, it suggested that producing genetically modified animals, as Wilmut had envisioned, was indeed possible. There was no reason, given the birth of Megan and Morag, that similar work could not be done using genetically modified cells grown in cell culture.

Table 1   Success rates in Dolly's cloning

	Number	Percent
Reconstructed eggs	277	100%
Reconstructed eggs recovered from oviducts	247	89%
Development to morula / blastocyst stage	29	10%
Pregnancies	1	<1%
Live lambs born	1	<1%

Given this success, the cloning of Dolly, using an adult donor cell, was really just icing on the cake for Wilmut and Campbell.[20]

The protocol for this experiment was nearly identical. The donor cells were found in the freezer at the Roslin Institute. They were mammary gland or udder cells that had been saved from a six-year-old ewe, whose fate was unknown. As in the previous experiments, these cells were grown in culture and then induced to enter the G_0 phase. They were then fused with enucleated eggs. Success rates were low. Out of 277 fusions, only one lamb was born.

However, one was enough. Dolly was born on 5 July 1996. She was named after the country singer Dolly Parton, partly to reflect her genetic origins as a mammary gland cell.

HOW DO SCIENTISTS KNOW A CLONE IS A CLONE?

You might be wondering just how scientists can be so sure that Dolly was a clone. In Dolly's case, only one animal survived out of 277 attempts. Maybe the surrogate ewe was already pregnant or perhaps Dolly resulted not from a differentiated cell but a stray immature cell accidentally picked up in the laboratory.

To show definitively that two animals are genetically identical, scientists use a technique known as DNA fingerprinting. This is the same method used in paternity tests and by police forces around the world to match evidence collected from crime scenes with particular suspects. It works by comparing highly variable DNA segments across samples. If a large number of typically variable segments match, scientists can conclude that the samples are genetically identical. Dolly's genetic fingerprint matched the donor cells precisely. Using statistics, the scientists concluded the probability that a second sheep would share, by chance, this same profile was about one in two billion. In short, it is more likely that you will win the lottery or get struck by lightning than it is that Dolly was born from a stray contaminating cell.

Scientists also use simple techniques to help verify that a clone is truly a clone. Dolly, for instance, was cloned from a Finn Dorset sheep, which has a white face but developed in a Scottish Blackface surrogate. Because Dolly had a white face, scientists know she was not the offspring of her surrogate.

Figure 12 Dolly and her surrogate mother. Note that the surrogate is a Scottish Blackface sheep while Dolly is a Finn Dorset and has a white face. (Image courtesy of Roslin Institute)

When Dolly's birth was announced the following February, it stunned the world. She made headlines from New York to New Zealand and everywhere in between. Scientifically, she was significant because she proved that cloning using a differentiated adult donor cell was possible. She answered, once and for all, the question Weismann had posed a century before. Her birth also set off a flurry of follow-up research, repeating and extending the technique. But Dolly's impact extended well beyond the scientific world. She gained lasting fame not because of the scientific doors she opened but because she brought the specter of human cloning once again to the fore.

Further reading

Gina Kolata's *Clone: The Road to Dolly and the Path Ahead* (William Morrow and Company, 1998) is an accessible and entertaining book-length treatment of the events recounted in this chapter. Kolata's interviews with many of the participants in the recent history of cloning paint a fascinating picture of how the field developed.

For more details on the actual cloning of Dolly and the work at Roslin Institute leading to her birth, *The Second Creation: Dolly and the Age of Biological Control* (Farrar Straus Giroux, 2000) is a good choice. Written by Wilmut and Campbell – two of Dolly's creators – and a science writer, it provides an insider's perspective on the research.

For those interested in scientific scandal, Rorvik's *In His Image: The Cloning of a Man* (J.B. Lippincott, 1978), though out of print, is readily available from used booksellers. Given the recent advancements in cloning, the science is rather outdated. Still, the story feels surprisingly contemporary.

4

Animal cloning in the twenty-first century

The announcement of Dolly's birth initiated a major change in cloning research. From the confines of the farm, cloning quickly made its way back into the mainstream of biomedical research. As early follow-up work proved successful, the research funding agencies that had previously scorned the field opened their coffers. Scientists quickly followed the money.

This excitement was not a surprise. Although long out of favor, cloning offered scientists a large number of opportunities. The technology provided an ideal means to study development. This, of course, is why Hans Spemann proposed cloning in the first place. It also opened the door to a number of commercial and medical applications, from the cloning of house pets to the deliberate creation of transgenic animals. Although few applications of cloning technology have come to market, research has advanced significantly in many of these areas. However, cloning remains inefficient and many challenges remain if the technology is to reach its full potential.

Cloning Noah's Ark

Following Dolly's birth, a top priority for scientists was extending the cloning technique to work with other species. Although most mammals are similar, and scientists often assume a technique will transfer from species to species, it was not certain that the protocol used to clone Dolly would work with other animals. Sheep seemed particularly amenable to cloning, while other animals, including mice, had proven more challenging. With great anticipation, scientists awaited the first reports of cloning from differentiated adult cells in other species.

The wait was not long. Before 1998 was over, two groups had cloned cattle from differentiated cells and the first extension of the technique to work with mice was reported. These results proved that Dolly was not a fluke and thus legitimized the field. Furthermore, the successful cloning of cows suggested that the vision of producing therapeutic compounds in milk might well be feasible, while the successful cloning of mice verified the utility of the technique as a basic biomedical research tool.

Successful cow cloning was reported by two groups in 1998. One group, led by Jose Cibelli at the University of Massachusetts, used the nuclear transfer technique to create transgenic cows. To accomplish this feat, the scientists used donor cells that had been genetically modified to express a bacterial gene. They created 276 reconstructed eggs using nuclear transfer and ended up with three healthy calves. These calves were all genetically identical and expressed the bacterial gene in their cells. This success rate represented an increase in efficiency over the cloning of Dolly, when 277 fused eggs were required to create a single healthy clone. The other confirmation that cows could be cloned came from a group of Japanese scientists. While their cloned cows were not transgenic, they were cloned from differentiated adult, rather than fetal, cells. They were also produced with relatively high efficiency. Notably, eight of the ten blastocyst stage embryos they transferred to surrogate mothers developed to term.

Mice are the workhorses of mammalian developmental biology labs. Over the years, scientists have devised a host of procedures for working with these small rodents and manipulating their genes. They have also bred countless varieties, or strains. Today, you can open a catalog and order a mouse tailored to answer a specific research question. Mice that are predisposed to tumor development, heart attacks, or insulin resistance are merely a phone call away. Thus, the successful cloning of mice was a critical step if the technology was to develop into a truly useful developmental biology research tool.

Although earlier attempts had proven unsuccessful, using the template provided by Dolly's cloning, scientists were able successfully to clone mice.[1] This groundbreaking research took place primarily at the University of Hawaii and was conducted by postdoctoral researcher Teruhiko Wakayama, laboratory director Ryuzo Yanagimachi, and several colleagues. Interestingly, the scientists returned to the injection method first used by Briggs and King to clone frogs nearly fifty years before instead of the fusion method used to clone Dolly. They produced more than fifty living cloned mice. These mice were derived from cumulus cells, a type of cell which surrounds and protects developing eggs, and so the first cloned mouse, born on 3 October 1997, was named Cumulina. She went on to produce two litters and lived for two years and seven months; seven months longer than the average mouse.

As well as cloning the first mice, these scientists also reported another significant advance. Because mice mature quickly, Wakayama and his collaborators were able not just to clone once but also to make clones of clones, marking the first time a cloned mammal had itself been cloned. A second-generation cloned mouse was genetically identical not just to its parent but also to its parent's parent, the original nuclear transfer donor. The second generation of cloned mice was produced at a similar efficiency level as the first and no differences were reported in the health of the two generations.

Following these two important advances, scientists continued to expand the range of animals that could be cloned. In 1999 came the first cloned goats: three healthy female goats were born following nuclear transfer from differentiated fetal cells. Just a year later, cloned pigs were reported. The successful cloning of pigs represented an important step for two reasons: first, pigs are an important food source in much of the world and livestock producers are interested in using cloning technology to increase the efficiency of meat production, and second, because many of their internal organs are similar to humans in size and function, pigs are considered an important potential source of organs for transplant. Cloning opens the door to creating special lines of pigs that are genetically modified and bred specifically for transplant.

More recently, scientists have successfully cloned cats, rats, rabbits, horses, and deer from differentiated cells. They have even cloned a mule – a sterile animal resulting from the mating of a male donkey with a female horse – showing that cloning can replicate animals that could not possibly breed naturally. In short, although some animals have proven more difficult to clone than others, scientists have been able successfully to clone most mammals with which they have experimented.

One challenge has been dogs. Dogs have an unusually complex reproductive system, in which eggs develop significantly after ovulation but before fertilization. This necessitates modification of the standard cloning procedure and has hindered attempts to clone this most popular of house pets. Finally in 2005, after several years of trying, one research group reported successfully cloning a dog.[2] This cloned dog, named Snuppy or "Seoul National University puppy" after the institution where it was created, quickly gained worldwide fame and is perhaps the second most famous cloned animal ever. Snuppy was even named the "2005 Invention of the Year" by *Time* magazine. Following reports of fraud in other cloning experiments carried out by the laboratory that created Snuppy (see Chapter 5 for more details), some scientists wondered if Snuppy was truly a clone. An investigation completed in early 2006 into Snuppy's pedigree verified that the dog was indeed a clone and deserves his place in history.

Another exception has been non-human primates. Despite numerous attempts, scientists have, for the most part, been unable to clone humankind's closest animal relatives using the nuclear transfer technique. Most scientists believe these difficulties will be overcome with time and that monkeys will eventually be cloned routinely like mice, cows, or sheep, but some are more skeptical. One group, led by Gerald Schatten at the University of Pittsburgh, attempted to clone rhesus monkeys, a species that, due to its physiological similarities with humans, is in great demand for medical research. Following unsuccessful attempts to clone these monkeys, Schatten's research group examined why the procedure was failing. They concluded that the enucleation process removed a specific protein crucial for the formation of the mitotic spindle and that, without this protein, normal development was not possible. Perhaps recalling the bold statement McGrath and Solter used to conclude their 1984 report, Schatten's report ended with the following statement, "With current approaches, nuclear transfer to produce embryonic stem cells in nonhuman primates may prove difficult – and reproductive cloning unachievable."[3]

All this experience has allowed animal cloning techniques to advance. You may recall that the donor cell used to create Dolly was induced to enter a resting state, G_0, before nuclear transfer was attempted. Scientists initially thought this step might be crucial to the success of the cloning process. Experience has shown, however, that while cloning is often more efficient when donor cells are in this resting state, it is not a requirement. Indeed, several cows have been cloned using donor cells in other phases of the cell cycle.

The health of cloned animals

Beyond expanding the range of animals that can be cloned, research has focused on attempting to improve the efficiency of the procedure and ensure the health of cloned animals. This remains a major goal of scientists today, as inefficiency in the process hinders essentially all of the potential applications of cloning.

Even if you know little about cloning, you may have heard about these concerns in the mass media. Dolly's health was a major news story; the BBC, CNN, and other outlets reported her various medical conditions. She developed arthritis in 2002, and was put down in 2003, at six years old, after being diagnosed with progressive lung disease. Six is young for a sheep to have developed these conditions and some believe they were associated with her cloning. However, scientists will tell you that while this is a possibility, no conclusions can be reached from a sample size of one. Larger controlled studies will be needed to understand what impact, if any, cloning has on the long-term health of cloned animals. This hasn't stopped anti-cloning activists, however, from claiming that cloning is inhumane and that it led to Dolly's early demise.

To understand better the health concerns associated with animal cloning, we will look at what scientists have found when they have examined not just one animal but the hundreds of animals that have been cloned in the last decade. The first thing that becomes clear is that we are asking two different questions. Most of the inefficiency of the cloning process is associated with embryos that fail to develop properly and are lost before birth. Health defects in surviving animals, on the other hand, affect only a fraction of animals that develop to term and survive their first few days of life. We will consider these two issues in turn.

Embryonic and fetal loss during cloning

In discussions about Dolly, the ratio 1:277 is often cited. This is indeed the overall success rate associated with the experiment. Of 277 fused eggs, one animal developed to term and survived birth, but this low success rate does not imply that the Roslin scientists created hundreds of sheep-like monsters without legs or with deformed hearts. Most of the difficulties occurred early in the process. Only twenty-nine (roughly 10 percent) of the 277 fused eggs reached the morula or blastocyst stage.[4] (Remember, these stages are early in development before the microscopic embryo has been transferred to a surrogate mother.) These twenty-nine embryos were the only ones of the original 277 that were transferred to the wombs of surrogate ewes. Many of these embryos either failed to implant or spontaneously aborted early in development; when the surrogates were examined by ultrasound around day fifty, only one was found to be pregnant. This pregnancy proceeded normally and led to the birth of Dolly. Thus, the 1:277 ratio is a bit of an over-simplification. It is true that cloning is inefficient but, at least in this case, most of the inefficiency happened early in development.

Similar stories of embryo loss have played out in numerous experiments. By comparing the efficiency of the cloning process across many experiments, scientists have begun to develop an idea of how the process typically works. Although there are a few exceptions, in general the overall efficiency is less than 5 percent: in 100 attempts to create a cloned animal, fewer than five animals develop to term and go on to live healthy lives. In some cases, like Dolly's, most of the embryo loss occurs during early development. In others, such as Wakayama's mouse experiments, this loss is more evenly distributed.

Although many of the embryos that are lost are never recovered, some are, and scientists have been studying them, in an attempt to understand what goes wrong during the development of cloned animals. Among embryos that implant and start to develop, the most consistently observed defects relate to placental development. There is a growing scientific consensus that typically, it is not the embryo proper that develops inappropriately but the extra-embryonic structures. The placenta plays a crucial role in nourishing developing fetuses; over time, placental abnormalities give rise to other developmental abnormalities and may lead to embryo loss. In many of the cases where specific abnormalities have been observed, scientists believe these defects are secondary to placental failures. This simply means that the defects were not caused by the cloning procedure directly but by the placenta I abnormalities associated with the cloning. Thus, if scientists could somehow figure out why the placenta develops incorrectly and correct this defect, many of the other abnormalities might be averted.

As the placenta doesn't develop before implantation, the high rates of embryo loss observed before development to the blastocyst stage cannot be due to poor placental development. In these early stages, success rates seem to depend on a host of factors, including the specific donor cell used for the nuclear transfer procedure and the specific experimental protocol used to remove the egg's genetic material and insert the donor genetic material. In Wakayama's mice experiments, for example, nearly sixty percent of cloned embryos developed to the morula or blastocyst stage when he used cumulus cells as the nuclear donors.[5] When he tried two other cell types, success rates dropped significantly: only twenty-two percent of cloned embryos created with brain cells and forty percent created with sertoli cells (a type of cell that nurtures sperm during development) reached these stages.

Some reconstructed embryos fail to develop due to damage caused by the enucleation and fusion procedures. Although identifying the causes of developmental failure in these early stage embryos is challenging, scientists are hopeful that the procedure can be optimized by comparing multiple experiments, eventually leading to a consensus about which cells and which experimental methods work best.

Taken together, early embryo losses and defects associated with placental development probably account for many of the cloned embryos lost during gestation but losses caused by other factors certainly also occur. Identifying specific defects and linking them to the cloning procedure has proven difficult, as many of the abnormalities observed vary by species, donor cell, and other factors. Scientists are at work on this problem and strategies to reduce gestational losses are under investigation. A key goal is to develop criteria to help predict which blastocysts will develop normally and which will not. Since the gestation of cloned embryos in surrogate mothers is quite costly, particularly in large mammals, such a method would greatly facilitate the potential commercial applications of cloning.

Although scientists are optimistic, conducting the large-scale studies required to understand gestational losses during cloning is an expensive and long-term project. For this reason, it may be some time before these losses are understood and scientists can devise strategies to improve the overall efficiency of the cloning process.

Health problems in surviving clones

In contrast to the somewhat dire image provided by examining embryo losses during the development of cloned animals, examination of the health of surviving animals paints a relatively more optimistic picture. It is true that some cloned animals are born with a wide range of abnormalities, but by and large animals that develop to term and survive a few days after birth go on to live reasonably healthy lives. One study examined 335 cloned animals from five different species and found that 259 (77 percent) of these animals were reported to be healthy.[6] The remaining 23 percent had some sort of developmental problem, ranging from kidney defects and hypertension to bacterial infection, reported after birth.

Some of these defects were similar to abnormalities that have been observed in animals produced by other assisted reproductive technologies. The most frequently noted of these is "large offspring syndrome," where animals are atypically large at birth and often have mal-developed organs and musculoskeletal systems. Because this and similar conditions are also seen during *in vitro* fertilization in these animals, the defects are not thought to be caused by the cloning technique directly but by the associated embryo manipulation.

Other defects observed in cloned animals are almost certainly attributable to the cloning procedure. These include obesity, which has been observed in a number of cloned mice, and hypertension, osteoporosis, and anemia, which have been reported in cloned cows. Most of these sorts of defects appear only irregularly, and often in a single species, challenging scientists hoping to understand how and why the cloning procedure works or fails. Data on the health of cloned animals are both promising and troublesome. The conclusion that more than three-quarters of cloned animals were healthy is promising, particularly considering the technology has been in use only a short period of time, but the finding that a quarter of surviving offspring exhibit a variety of abnormalities is far from ideal. This error rate is certainly higher than that associated with normal reproduction or other assisted reproductive technologies, suggesting that progress still needs to be made.

One group of scientists, which includes the cloning pioneer Ian Wilmut, has conducted detailed pathological studies on cloned lambs to test the preliminary conclusion that most cloned animals are seemingly healthy. They examined eight sheep that died at birth or shortly afterward and found a variety of defects, many of which offered parallels to rare genetic human disorders. These findings call into question the idea that most animals that develop successfully to term are healthy. Such detailed pathological studies cannot be conducted on true surviving clones, however, as many of these animals are still living.

An important, and less controversial, finding is that the offspring of cloned animals are healthy. Even when born to clones with clear abnormalities, such as obesity, the offspring almost without exception do not exhibit their parents' abnormalities. This is crucial for those hoping to commercialize cloning technology. While cloned animals may not be considered safe for a variety of uses, it appears that their offspring are. While it might be some time before your local supermarket stocks milk from cloned cows, milk from their offspring might be available soon.

At first glance, the finding that the offspring of cloned animals are healthy might seem surprising but to scientists studying cloning, it is not particularly so. To see why, we need to understand that the changes that affect development in cloned animals are not genetic but "epigenetic" in nature. Epigenetic changes are those that affect gene expression without affecting the underlying DNA sequence. Developmental errors that result from faulty reprogramming during the cloning process are almost exclusively epigenetic. The DNA sequence is correct, but the process of reading genes and producing proteins is not working quite properly. When a cloned animal gives birth to an offspring produced by normal fertilization, the epigenetic signals governing expression in the offspring are reset and the offspring develops normally.

EPIGENETICS AND CLONING

Many of the differences in gene expression between mature differentiated cells and embryonic cells are due to epigenetic changes. These are alterations made to chromosomal proteins or to a DNA molecule's structure without altering its sequence. When an animal is cloned from a differentiated cell, the donor nucleus contains these epigenetic changes and needs to be reprogrammed. The birth of Dolly, and countless other cloned animals, proves that reprogramming is possible but scientists believe it works only rarely. Reprogramming failures account for some portion of the low success rate observed in cloning studies.

Some of the most striking evidence to support this theory comes from studies of mice. A group of scientists, led by Rudolf Jaenisch at the Whitehead Institute for Biomedical Research, compared the expression of ten thousand genes in cloned and normal mice and found that 4 percent of genes were abnormally expressed in the placenta of cloned animals.[7] Differences between cloned and normal mice have also been found when scientists examined imprinting and DNA methylation, two important epigenetic modifications that play roles in development.

EPIGENETICS AND CLONING (*cont.*)

The epigenetic reprogramming that allows a differentiated nucleus to direct embryonic development remains poorly understood. Proteins found in the egg's cytoplasm are thought to play key roles but which proteins matter and how they work are open questions. One goal for this research is the development of protocols that make reprogramming more efficient. If such protocols are devised, they may well increase the efficiency of the cloning process.

Applications of cloning technology

Although inefficiency continues to hinder the application of cloning technology, scientists have made progress. Most of these successes are proof-of-principle experiments that show a given application may be feasible in the long run, assuming both efficiency and health concerns are overcome. These applications fall roughly into three categories: first, the cloning of valuable animals, including endangered species, pets, and commercially valuable animals; second, the use of cloning to create transgenic animals, for a variety of reasons including improving livestock species and producing pharmaceuticals; third, the use of cloning to create human embryonic stem cells, which could theoretically be used in medical therapies – the so-called therapeutic cloning paradigm. Discussion of therapeutic cloning will be deferred until Chapter 5, where human embryonic stem cells are examined. The remainder of this chapter will examine the other two groups of applications.

Cloning valuable animals

The cloning of commercially valuable animals is among the technology's least controversial and most advanced uses. Animal breeders seeking to produce high quality animals already frequently use assisted reproductive technologies, including sperm freezing and artificial insemination. To many in the animal husbandry industry, it is just a short step further to use cloning technology to produce genetic replicas of top animals.

As long as cloning remains as inefficient as it is today, it cannot provide a cost-effective means of mass-producing animals; at the moment, it remains an option only for a small number of exceptionally valuable individuals. However, some industry insiders expect cloning technology to take off rapidly in the United States if the government allows products derived from cloned animals, or products derived from their offspring, to enter the market. Such a prediction shows confidence that efficiency can be improved, at least for commercially important species such as cows and pigs.

Cloning for conservation

A more controversial application is the use of cloning in the conservation of endangered species. Nuclear transfer technology may provide a means of promoting the survival of species nearing extinction or, in rare cases, perhaps allow scientists to bring back recently extinct species. With nearly 7,300 animals species listed as "threatened" in the 2004 report of the Species Survival Commission, the appeal of such a strategy is obvious.[8] Attempts to clone endangered species, such as the giant panda, and extinct species including the Tasmanian tiger, have generated significant media attention, despite questionable prospects of success.

Cloning endangered species poses many scientific challenges and cloning extinct species countless more. Little, if anything, is known about the reproductive physiology of most endangered species. Sheep had been studied for many years before they were successfully cloned from adult cells and the same can be said for mice, cows, and many other animals. In addition some animals almost certainly died as scientists perfected these techniques. In contrast, few scientists study most endangered species and these rare and threatened animals cannot be sacrificed in attempts to learn more about their physiology.

Even if, to take an example, experts in the reproductive system of the endangered snow leopard could be found, the availability of eggs for nuclear transfer would pose another problem. Assuming a cloning efficiency of 5 percent, the high end of most present estimates, it would take twenty eggs to create a single cloned animal. More likely, given our limited knowledge of reproduction in these species and the lower efficiencies typically seen in the first attempts to clone a species, successfully cloning an endangered animal would require a hundred or more unfertilized eggs. It is not clear where these eggs would come from. Unlike cows, whose ovarian tissue is readily available at the slaughterhouse, or pets, whose eggs can be obtained when animals are spayed, obtaining the eggs of an endangered animal would probably require surgery. The risk such an operation would pose to the animal might not be justified by the small chance that a healthy clone would be born.

Ironically, given that most people thinking of conservation efforts focus on large mammals, like the giant panda that is the logo of the World Wildlife Fund, the challenges associated with cloning endangered species suggest that other animals might provide better opportunities. Among mammals, litter-bearing species, such as rodents, offer the advantage of being able to transplant many embryos to a single surrogate. Amphibians and fish are perhaps even more promising. Many of these animals reproduce using external fertilization and lay hundreds or thousands of eggs. Thus it may be easier for scientists to obtain the necessary raw material to conduct experiments without exposing the threatened animals to unnecessary risks.

Given these complications, particularly of cloning large endangered mammals, you might be asking how the Chinese government plans to clone the giant panda? And what about media reports of the successful cloning of an endangered guar – a wild ox native to Southeast Asia? These, and similar cloning attempts, sidestep the difficulties by using a variant of the standard cloning procedure: "interspecies nuclear transfer."

In this variant, animals from two different species are used. The donor cell is taken from the endangered species but instead of fusing this cell with an egg from the same species, scientists fuse the donor cell with an egg from a closely related but not endangered species. In the case of the guar, which was the first to be reported, scientists used cow eggs.[9] They also used cows as the surrogate mothers. The cloned embryo that resulted from this nuclear transfer contained nuclear DNA from a guar and mtDNA from a cow. It was not clear if such a hybrid embryo could develop to term but, probably because of the close genetic relationship between the two species, one did and was apparently healthy at its birth in January 2001. Unfortunately, this guar, Noah, contracted a bacterial infection and died two days after birth. It was unclear if the illness was related to the cloning but such infections are not uncommon in newborn calves.

Greater success was reported later in 2001 by a group of European scientists working mostly in Italy. Using a similar technique, they successfully cloned an endangered mouflon, a wild sheep found on the islands of Sardinia, Corsica, and Cyprus.[10] Donor cells were isolated from two animals found dead in a pasture and injected into enucleated sheep eggs. Sheep, related to mouflons, were also used as surrogates. The research group started with twenty-three reconstructed eggs. Seven reached the blastocyst stage, leading to two pregnancies and one surviving animal. The cloned mouflon was reportedly healthy at seven months of age but went on to die from pneumonia relatively early in its life.

Since these first two reports, the technique has gradually been extended to other species. The group that cloned the guar has been granted permission to attempt to clone a bucardo, an extinct mountain goat found in Spain. The last bucardo died in Spain in January 2000 when its skull was crushed by a falling tree. Scientists preserved its genetic material and are hopeful it can be cloned. The same group also cloned two bantengs, a species of endangered Asian cattle. One of the two clones was abnormally large at birth and was put down but the other remains healthy and, as of June 2006, was on display at the San Diego Zoo. Other scientists have cloned African wildcats using domestic cats as the egg donor and surrogate. Although these wildcats are not endangered, the scientists believe their work with the interspecies transfer technique will facilitate the cloning of other endangered carnivores.

Because only a thousand or so giant pandas – one of the most recognizable symbols of China – remain in the wild, the Chinese government has expressed interest in cloning this species. Chinese officials have been using various assisted reproductive strategies for many years in an attempt to save the panda from extinction and their interest in cloning is a continuation of this strategy. Unfortunately for the scientists at the Chinese Academy of Sciences who are leading this effort, there are no well-studied close relatives of the giant panda to use as the egg donor and surrogate. Scientists have created panda embryos through fusion with both rabbit and cat cells and implanted a panda embryo into the uterus of a cat in 2002. The cat died two months later but research is continuing. Scientists are currently searching for more suitable surrogates.

A similar but perhaps even more ambitious project to clone the extinct Tasmanian tiger, a wolf-like marsupial, was undertaken by the Australian Museum. The last Tasmanian tiger died in captivity in 1936. However, one 139-year-old pup was preserved in pure alcohol and scientists were initially optimistic that cells could be isolated from it. Despite these hopes, scientists found the available DNA was badly degraded and the project was abandoned in early 2005. This failed attempt illustrates one of the major challenges associated with cloning extinct animals: suitable cell samples are unlikely to be available.

Scientists pursuing interspecies nuclear transfer of endangered species do not currently envision introducing these cloned animals to the wild. Rather, they are intended to maintain or increase the breeding population in captivity. Because these cloned animals are hybrids, with their nuclear and mitochondrial DNA derived from different species, it is not even clear if they should be considered the same species as the endangered animals that provided their nuclear DNA. However, this uncertainty need not affect their offspring. If a cloned male were bred with a naturally conceived female, the offspring would possess both nuclear and mitochondrial DNA from the endangered species (because mitochondria are inherited exclusively from the mother). Using this strategy, interspecies nuclear transfer could provide a useful tool to propagate the DNA of male animals, particularly if scientists hope to maintain genetic diversity in a small population containing one or more infertile males.

While scientists are progressing toward this vision of cloning endangered species, and have achieved some success with interspecies nuclear transfer, a bigger question has been raised. Is this research even a good idea? This is far from certain. Cloning has splintered the conservation community. Some believe it is useful as a last resort. Others are adamant that the attention given to the cloning of endangered animals is counterproductive. What good is it, they ask, to spend millions of dollars cloning an endangered animal when its natural habitat is being destroyed at ever-increasing rates? It would be better, in their opinion, to focus on more traditional conservation goals, including habitat preservation and protecting living animals from poaching or other dangers.

Cloning Fido or Spot

While cloning for conservation remains under development, another controversial use of the technology has reached the market: the cloning of domestic pets. Although some people consider this use frivolous, many pet owners develop deep feelings for their animals and have expressed an interest in cloning their cats or dogs.

In the United States in the last few years a number of companies, including Genetic Savings & Clone and PerPETuate, formed to offer pet tissue preservation and cloning services. These, and similar firms, offer tissue banking for pet owners hoping to clone a pet in the future, when the technology might perhaps be both cheaper and more reliable. Typically, customers pay an upfront fee of $700 or more and an annual fee of approximately $100 for storage of the cell samples. Genetic Savings & Clone, before going out of business in late 2006, also offered cat cloning. In early 2006, the company's website reported that the firm had cloned six cats, two of which were delivered to paying customers. The cost of a cloned cat, guaranteed to physically resemble its genetic parent, was $32,000. The company had hoped this price would fall as cloning technology advanced and planned to offer dog cloning once that technology was developed and refined.

Pet cloning has drawn criticism from a variety of groups including the Humane Society of the United States. In a 2002 press release following the announcement of the first cloned cat, this organization wrote: "it doesn't sit well with us to create animals through such extreme and experimental means when there are so many animals desperate for homes."[11] In the United States, an estimated three to four million cats and dogs are put down in animal shelters each year.[12] However, while the fate of these animals is unfortunate, proponents of pet cloning argue that the small number of cloned pets has essentially no impact on the pet overpopulation problem.

Another argument voiced in opposition to pet cloning is that the practice is potentially deceptive. Desperate pet owners, these critics argue, are tricked into paying thousands of dollars for a pet that may or may not be similar to their lost animal. Pet owners may be convinced, or convince themselves, that genes determine everything about their pet, ignoring the role played by the environment. The first cloned cat, CC or CopyCat, looked quite different from her genetic donor, because her parent was a calico (tortoiseshell) cat; a species whose distinctive coloration and pattern are determined randomly during development. This is an exception; cat coloration is normally largely genetically determined. This is not the case for all traits, however, particularly those affecting behavior or demeanor. It is not clear that a cloned pet would behave like its genetic donor. Genetic determinism – the fallacy that genes control all aspects of a living organism – will be examined in more detail in Chapter 6, when we look at the debate over human cloning. Suffice it to say that it could potentially confuse pet owners considering cloning. However, if pet cloning companies accurately explain the technology and pet owners considering cloning understand they would be paying for a pet likely to be similar in some ways and different in others from their original pet, this concern could perhaps be averted.

Although demand for cat cloning was not sufficient to keep Genetic Savings & Clone in business, if cloning efficiency increases and prices go down, other companies seem likely to fill this market niche, at least in the United States. Less progress and, indeed, less interest in the technology has been seen thus far in other countries. If pet cloning emerges as a viable business in the United States, however, there is no reason why pet owners around the world couldn't take advantage of these services.

Creating transgenic clones

As we discussed briefly above, the primary reason Ian Wilmut and Keith Campbell cloned Dolly was to develop a method for creating transgenic livestock. In the years since Dolly's birth, this area of research has advanced significantly, with numerous groups inserting a variety of genes into cloned animals. This technique combines cloning with genetic engineering and, not surprisingly, is controversial. It promises a wide range of potential benefits, including the production of mad cow disease resistant cattle, the generation of animal organs suitable for transplant into humans, and the efficient production of therapeutic proteins for use as medicine.

Livestock producers are one of the groups interested in using cloning to generate transgenic animals. Their interest derives from the potential of the technology to increase the efficiency of the animal production process. Although many increases in efficiency could be imagined, time-to-market and food consumption are important concerns for animal producers. Transgenic animals that grow more quickly or consume less food per kilogram of growth would be particularly valuable. One strategy is to introduce genes that promote growth, for example those coding for growth hormones, into important livestock species, such as cows and pigs. Overexpression of growth hormone in this manner does increase growth but thus far has been accompanied by undesirable side effects. Another strategy is to increase the quantity and quality of food produced by the animals themselves. For instance, in swine production, the quantity of milk produced by a lactating sow is a limiting factor for piglet growth. Scientists are exploring ways to use cloning technology to create pigs that produce more milk: one estimate suggests that a 10 percent increase in milk production by lactating sows would lead to $28.4 million annually in additional earnings for U.S. pork producers.[13]

Improving the health of livestock is another possible use. Given the significant disruption that bovine spongiform encephalopathy (BSE), commonly known as mad cow disease, has caused the cattle industry worldwide, one active area of research has been the development of BSE-resistant cows. BSE and similar diseases (including scrapie, which affects sheep and goats) are caused by a single protein, known as a prion, which can take one of two forms. In its normal form, this protein is not harmful but converted to its pathogenic form, either by mutation or infection, the prion serves as a template for converting additional copies to the pathogenic form. Eventually, plaques of these prions form and disease results. Scientists are not sure exactly what role the normal prion plays in the cell but no abnormalities were detected when transgenic mice were created without it. Furthermore, these animals appeared resistant to the mouse equivalent of BSE.

These results raise the hope that, using cloning technology, transgenic sheep and cows could be created that would not be susceptible to prion diseases. In 2001, scientists first reported creating cloned sheep in which each cell had one of its two copies of the prion gene eliminated. Although four of these sheep were live-born, all died within twelve days of birth. The researchers do not believe this high mortality was related to the specific gene eliminated from the sheep but suggest that the prolonged cell culture associated with the genetic modification procedure may reduce the viability of these cells for the cloning procedure. A different strategy was used to create BSE-resistant cows in 2003. Rather than attempting to eliminate the prion protein, scientists created animals that over-expressed a special variant of the prion. This variant functions similarly to the normal prion but resists conversion to the pathogenic form. These variant prions block the formation of plaques and seem to halt the development of prion diseases. Four cloned calves expressing this variant were born in November 2003 and sent to Japan in 2004 to undergo tests to verify their disease-resistant status. If the results are promising, cattle derived in a similar manner may eventually find their way into the world's food supply.

Cloning technology may also one day play a role in human organ transplants. As medical technology advances, transplantation has become an option for an increasing number of patients suffering from late-stage organ failure. Unfortunately the demand for donated organs greatly outstrips supply and every year thousands of people die waiting for a suitable organ. One strategy to overcome the shortage of donated human organs is to use organs from other animal species, a technique called "xenotransplantation." Pigs, because many of their major organs are physiologically similar to humans', are viewed as the most promising species for this technology. Unfortunately, transplanting an organ from one species to another typically generates a severe immune reaction. In human–pig xenotransplantation this response is called "hyperacute rejection" and relies on the recognition of specific compounds, found on the pig cells by the human immune system. Scientists have identified the protein that is responsible for

XENOTRANSPLANTATION

Anecdotal reports documenting the transplantation of animal tissue to humans go back hundreds of years but only limited progress was made until the 1960s, when early immunosuppressive treatments became available. A 23-year-old schoolteacher survived for nine months after receiving a chimpanzee kidney in 1964, proving long-term survival following xenotransplantation was possible. Immune rejection has plagued this research, however, and nine months remains the record for the longest survival following complete organ xenotransplantation.

While organ transplantation has proved difficult, less complex procedures have been more successful. Heart valve transplants, typically from pigs, have been a particular success: heart valves from pigs have been transplanted successfully into hundreds of thousands of patients over the last thirty or so years.

Improved immunosuppressive drugs and the creation of transgenic pigs using cloning technology are inspiring new interest in whole-organ xenotransplantation. Scientists are optimistic that organs from transgenic pigs will produce less severe immune responses and improve transplant success rates. Pig to human xenotransplantation still poses many challenges, however, and significant hurdles must be overcome before such transplants could be attempted.

synthesizing this compound and have created cloned pigs in which the gene coding for this protein is disabled. The next step in this research is to transplant organs from these pigs into non-human primates to assess their potential for use in humans.

Scientists and private companies remain interested in using cloning to create transgenic animals that produce pharmaceuticals but apart from a few proof of principle experiments, progress toward this vision remains limited. The most prominent proof of principle came from scientists at the Roslin Institute. After creating Dolly, they extended their work to create cloned sheep that produced a useful therapeutic compound in their milk. These sheep were produced using donor cells that had been genetically modified to include a gene coding for human clotting factor IX.[14] Two of these sheep, Polly and Molly, survived. The protein they express plays a key role in blood coagulation and its deficiency results in the blood disease hemophilia ß. This illness is currently treated with protein isolated from human plasma. Theoretically, isolating this protein from animal milk could be more economical.

Because most of the work takes place in private companies and is kept secret pending patent application, it is hard to assess how far scientists have advanced in this area but a number of companies have expressed interest in developing the technology. Pharmaceutical production is tightly regulated in many countries and regulatory acceptance of proteins produced in cloned animals is a key issue that must be addressed before any of these compounds could be brought to market.

While these regulatory hurdles are navigated, scientists are exploring the best way to produce and isolate proteins from cloned animals. Producing proteins in milk has been the most commonly mentioned approach and is relatively well-understood. Yields of up to several grams of protein per liter of milk have been reported. However, other strategies are also under consideration. These include isolating proteins from urine, which may offer some advantages over milk. One key benefit is that urine is produced by both males and females of all ages. Milk, in contrast, is produced only by female animals that have reached sexual maturity. Other options include isolating therapeutic proteins from blood or seminal fluid. Pig seminal fluid is considered promising because of its high protein content. Little is known about the control of gene expression in this fluid, however, limiting these prospects in the short-term.

When or if products from transgenic cloned animals will make their way to market is anyone's guess. However, because many of these technologies appear commercially attractive, research is likely to continue. Only time will tell if herds of cloned cattle will one day replace pharmaceutical manufacturing facilities.

5
Embryonic stem cells and the promise of therapeutic cloning

When Dolly was first introduced to the world, she opened the door to the potential cloning of humans yet few were truly interested in taking this step. Although some publicity-seekers claimed to be cloning humans, the vast majority of scientists denounced the practice. This debate was changed forever by the successful isolation of human embryonic stem cells in 1998. Following this breakthrough, most scientists continue to denounce the cloning of humans for reproductive purposes but many embrace the concept of therapeutic cloning – the creation of cloned human embryos for the purpose of deriving human embryonic stem cell lines. The hope that this still-hypothetical possibility inspires is one of the driving forces behind the furor over human embryonic stem cell research in the United States and around the world.

In the, still-theoretical, therapeutic cloning process, an adult human cell is used as a donor for cloning and the resulting cloned human embryo is allowed to develop until the blastocyst stage, when human embryonic stem cells are isolated. These cells can develop into any of the many types of human cells and this unique ability gives them significant therapeutic potential. Furthermore, because the cells are genetically identical to the patient, the risk of immune rejection following transplant is greatly reduced, if not entirely eliminated. This combination of benefits has doctors, scientists, and patient advocates excited about the long-term

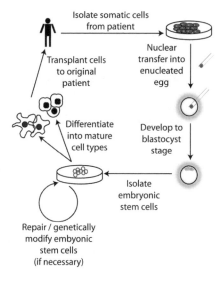

Figure 13 A general outline of the therapeutic cloning protocol.

potential of human embryonic stem cells and therapeutic cloning. The research is controversial, however, because it involves the deliberate creation of a human embryo, which after the isolation of embryonic stem cells is typically no longer viable. This chapter explores the progress scientists have made toward this vision of regenerative medicine and the hurdles that remain.

The body's master cells

Stem cells are the body's cellular repair mechanism. They are a specialized class of undifferentiated, or partly differentiated, cells, whose role is to replenish the population of mature differentiated cells. Many stem cells are partially differentiated and under normal circumstances give rise to only a small subset of differentiated cell

THERAPEUTIC CLONING TERMINOLOGY

In recent years, as political debates over embryonic stem cell research have intensified, a debate has emerged over the terminology used to describe the derivation of human embryonic stem cell lines from cloned human embryos. The term "therapeutic cloning" will be used throughout this book, as this is the most common terminology outside the scientific literature and reflects the long-term goal of the research. Some scientists, ethicists, and policymakers prefer the term "research cloning" believing it more accurately describes the present state of research and reflects the reality that the creation of cloned embryos today is not for current therapies but for research that may lead to therapies in the future. Terminology matters, because it can affect public perception. The optimistic connotation of "therapeutic cloning" might encourage public support for the technology while the term "research cloning" might not.

types. Embryonic stem cells, in contrast, are a special class of stem cells, found only in the inner cell mass of developing embryos. They are undifferentiated and can give rise to all the cell types of a mature organism.

Scientists refer to this ability of one cell type to give rise to any other type as "totipotency" and the ability of a cell type to give rise to many, but not all, other cell types as "pluripotency." Scientists typically classify human embryonic stem cells as pluripotent, since in normal development they do not give rise to the placenta or other extra–embryonic cells.

During normal development, embryonic stem cells quickly differentiate and lose their trademark developmental flexibility. However, it is possible to isolate these cells from developing embryos and grow them in culture. These cultured embryonic stem cells can, if carefully tended, remain in their undifferentiated state almost indefinitely, allowing scientists to grow large numbers of them; a key step toward the development of any stem cell based therapies.

Growing stem cells in a dish

When human embryonic stem cells are isolated from a developing embryo and grown successfully outside the body, they are called a "human embryonic stem cell line." A single cell line can live for many years and be used for hundreds, if not thousands, of experiments. The first human embryonic stem cell lines were successfully isolated at the University of Wisconsin in 1998 and are the subject of numerous peer-reviewed research papers. Not all cell lines are stem cell lines: perhaps the most famous cell line of all time is the HeLa cancer cell line. It was isolated in 1951 from the cervical carcinoma of a patient named Henrietta Lacks and has been used in thousands of experiments.

Although scientists have long known that human embryonic stem cells exist, the successful isolation of a human embryonic stem cell line posed significant challenges. Scientists needed to determine at exactly what point the cells should be isolated from developing embryos, how to isolate them without damage and finally how to maintain them in cell culture in their undifferentiated state. Embryonic stem cells were successfully isolated from mice in 1981; it took seventeen years to replicate the process for human cells.

A major factor in the delay was the limited availability of human embryos on which to experiment. While mouse embryos are easy to acquire – they can simply be flushed from the oviducts of pregnant mice – human embryos are harder to come by and their use in research is controversial. Critics of research on human embryos point out that these embryos, if given the appropriate environment, might develop into healthy human beings and thus liken embryo research to abortion. We'll examine this ethical controversy in more detail later but suffice it to say that the controversial status of research on human embryos slowed the process of isolating human embryonic stem cells.

The first major step toward the isolation of these cells was reported by a team of scientists, led by Ariff Bongso, at Singapore's

National University Hospital. Bongso, a Sri Lankan who trained in Canada before settling in Singapore, is an expert in the field of *in vitro* fertilization and was a member of the team that produced Asia's first test-tube baby in 1983. One of Bongso's key contributions was the development of an improved culture system for human embryos. This system, sequential co-culture, worked by closely mimicking the normal developmental environment and was designed to improve *in vitro* fertilization success rates. A by-product of this new culture system was that human embryos could now be grown *in vitro* until day five, and in some cases day six, of development. At this point, human embryonic stem cells exist in the inner cell mass of the developing embryo and it becomes possible to think about isolating them. Bongso and his collaborators took one step toward this goal when they isolated inner cell masses from nineteen spare embryos donated by patients undergoing *in vitro* fertilization. These cells divided in culture successfully and initially remained undifferentiated. They did not remain undifferentiated for long, however, and the 1994 report describing this research did not receive significant attention.

The same cannot be said of James Thomson's work at the University of Wisconsin. When Thomson and his colleagues published their paper describing human embryonic stem cell lines in 1998, other scientists and the media quickly grasped its importance. Some of the excitement was due to the clear link between cloning and human embryonic stem cells. The stem cell report and the idea that cloning could be used to create embryonic stem cells once again brought the specter of human cloning to the forefront of public debate. Now, however, the debate had changed and respected scientists were supporting cloning to create embryos for medical purposes.

Thomson had a background in veterinary science and worked with non-human primate cells before switching to human cells. His initial stem cell success, reported in 1995, was the first successful isolation of a primate embryonic stem cell line. To accomplish this, Thomson started with an embryo flushed from the uterus of a fifteen-year-old rhesus monkey. He separated the inner cell mass from the surrounding cells using a technique known as "immuno-surgery." In this method, developed by Davor Solter in 1975, the trophectoderm cells surrounding the inner cell mass are selectively destroyed leaving the inner cell mass, and the embryonic stem cells it contains, intact. Thomson transferred these remaining cells to a new petri dish and placed them on a layer of inactivated embryonic mouse cells. These mouse cells, which scientists call a feeder layer, had been exposed to radiation and were no longer able to divide but the variety of small molecules and other nutrients they contained played a crucial role in allowing the embryonic stem cells to grow and divide without differentiating.

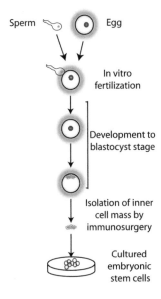

Figure 14 The protocol used to derive human embryonic stem cell lines from *in vitro* fertilized embryos.

Thomson used essentially the same technique to isolate human embryonic stem cells a few years later. Acquiring human embryos for the experiment was a key challenge, as flushing embryos from pregnant women is not considered ethically acceptable. Thomson, like Bongso before him, was able to use embryos created, but no longer needed, for fertility treatment. These embryos came from two fertility clinics, one in Wisconsin and one six thousand miles away in Israel.

Funding was also an issue. Thomson's research was funded by Geron Corporation, a publicly traded company based in California, rather than by the U.S. Government, the funding source for most basic biomedical research at academic institutions in the United States. Thomson turned to Geron for funding because the U.S. National Institutes of Health is prohibited from funding any research in which human embryos are destroyed. This restriction dates to a 1995 law but continues a longer tradition of U.S. taxpayers' dollars not supporting embryo research due to fears such research might encourage abortion.

With embryos available and funding in place, Thomson went ahead with the project.[1] He grew the embryos to the blastocyst stage *in vitro*, using the sequential co-culture technique, and isolated the inner cell mass through immunosurgery. As before, the inner cell masses were grown on inactivated mouse cells. Thomson and his collaborators started with thirty-six donated human embryos, of which twenty developed to the blastocyst stage. The inner cell mass was isolated from fourteen of these blastocysts and five distinct human embryonic stem cell lines were derived. Each of these lines came from a different embryo and could be maintained in its undifferentiated state in cell culture.

The availability of embryos for research

Because of the controversy surrounding human embryonic stem cells, considerable attention has been given to the question of

HOW DO SCIENTISTS KNOW CELLS ARE REALLY EMBRYONIC STEM CELLS?

Many cells look alike and scientists like Thomson must take steps to show that the cells they isolate are truly embryonic stem cells rather than similar-looking but less flexible cells. Several methods exist to make this case. One approach is to show that the cells can give rise to cancerous growths, called teratomas, when transplanted into immune-deficient mice. Teratomas – literally "monster tumors" – contain a wide variety of differentiated cell types and their formation is taken as evidence of pluripotency.

Embryonic stem cells will also spontaneously differentiate into spherical structures, known as embryoid bodies, in certain culture conditions. If scientists can verify that cells from all three primary tissue layers are present in an embryoid body, this is also taken as evidence of pluripotency.

The most stringent method to verify pluripotency is known as tetraploid embryo complementation. In this procedure, putative embryonic stem cells are injected into a blastocyst-stage embryo that has been modified so that it cannot develop on its own. When pluripotent embryonic stem cells are injected near the inner cell mass, they rescue this embryo and normal development can occur. In this case, the organism is derived entirely from the injected cells, while the extra-embryonic membranes are derived from the original modified embryo. This technique conclusively proves that cells are pluripotent, by showing that they can give rise to an entire organism. It cannot, however, for obvious ethical reasons, be used with human cell lines.

which embryos, if any, are appropriate for use in this type of research. While some people believe that human embryos should never be used in research, others make ethical distinctions between various sources.

By far the most commonly used source of embryos for human embryonic stem cell research is that used by both Bongso and Thomson: extra, or spare, embryos created during fertility treatment. These embryos are a by-product of clinics seeking to maximize the chance that a couple will successfully have a child. Women of reproductive age usually produce one mature egg a month but women undergoing *in vitro* fertilization are given hormone injections to induce the production of perhaps ten or more mature eggs. These eggs are fertilized and allowed to begin developing. Eventually a few are transferred to the woman's uterus. The remaining embryos are frozen for use if the first transfer procedure is unsuccessful or if the couple decides to have more children in the future. Unused embryos can remain frozen at fertility clinics long after parents have decided they do not want any additional children. Although these embryos could theoretically be donated to other couples, a practice known as embryo adoption, this is rare and the overwhelming majority of these frozen embryos are destined to be discarded. Because most of these extra or spare embryos are likely to be discarded, they are the least controversial source of embryos for human embryonic stem cell research. Indeed, some countries allow such research only with this source. The exact number of frozen embryos stored at fertility clinics and potentially available for research is unknown. One assessment estimated that approximately 400,000 such embryos were frozen in U.S. fertility clinics as of 2002.[2] Most of these embryos were destined for patient use but approximately 9,000 were pending destruction and another 11,000 were pending donation for research.

It is, of course, also possible to derive human embryonic stem cell lines from embryos created through *in vitro* fertilization for the sole purpose of research. While many people find the use of spare embryos created originally for clinical purposes ethically acceptable, the creation of embryos specifically for research is often seen as more problematic. Cells quite similar to human embryonic stem cells can also be isolated from aborted fetal tissue. These cells, called embryonic germ cells, were first isolated by John Gearhart and a team of scientists at Johns Hopkins

University.[3] Human embryonic germ cells share many important properties of human embryonic stem cells and could potentially be used in some of the same ways. Furthermore, research using fetal tissue, when this tissue comes from embryos aborted for reasons unrelated to the research, is often considered ethically acceptable. However, fetal tissue is difficult to acquire and remains somewhat controversial, due to its link with abortion. While the successful isolation of human embryonic germ cells was reported only days after Thomson's stem cell report, embryonic germ cells have thus far generated less excitement and follow-up research.

The last, and most controversial, source of embryos is cloning. Although at the time of writing this approach has never been reproducibly used to generate a human embryonic stem cell line, the theory is clear. Scientists hope to use somatic cell nuclear transfer to create a cloned human embryo that will develop successfully to the blastocyst stage. They then hope to isolate embryonic stem cells from this embryo. This research is extremely controversial because it opens the door to human reproductive cloning. Once scientists can grow cloned human embryos to the blastocyst stage, there will be no technical hurdles preventing them, or others who can replicate the technique, from transferring cloned human embryos to surrogate mothers, perhaps leading to the first cloned human. Some opponents of cloning embryos to derive embryonic stem cell lines argue that we should not take any steps that would facilitate the reproductive cloning of humans. But supporters of the research see the potential of therapeutic cloning as too medically promising to ignore. This combination of high hopes and deep fears makes the use of cloned embryos highly controversial.

Side-stepping immune rejection

Since all human embryonic stem cells have the theoretical potential to develop into any cell type, you might be wondering why cloned embryos are the most medically promising. The answer is simple but important: stem cell lines from cloned embryos may be able to avert immune rejection, a serious problem with transplant therapies. Rejection occurs when the immune system recognizes transplanted material as foreign and mobilizes to attack it. This attack may be rapid and strong – acute rejection – or it may be milder and persist for longer – chronic rejection. Either can lead to destruction of the transplant.

Transplant therapies have improved dramatically over the last half century but immune rejection remains a challenge. A key advance was the development of immunosuppressive drugs such as Cyclosporin, which was introduced in 1978. It, and similar drugs developed more recently, reduce this prevalence but at a cost. Transplant patients typically must take these drugs for life and suffer unpleasant side effects. In addition, suppressing the immune system increases a patient's risk of developing other infections.

The use of cloned embryos to create embryonic stem cell lines may overcome these immune rejection complications and simplify transplant therapy. If the transplanted material is genetically identical to the host, the immune system should not recognize it as foreign and thus immune rejection should not occur. The use of cloning by somatic cell nuclear transfer, if the technique is perfected for use with humans, should allow the development of embryonic stem cell lines genetically identical to patients. From there, scientists hope to direct the differentiation of these patient-matched cells into specific cell types that may be useful in therapies. Finally, mature differentiated cells would be transplanted to the patient.

Depending on the particular condition, scientists might want to combine this therapeutic cloning procedure with genetic modification. Embryonic stem cells, like other cells grown in culture systems, can be selectively modified. To treat patients suffering from genetic defects, it might be possible to derive an embryonic stem cell line from a cloned embryo, correct the genetic defect and transplant the cells back into the patient. This procedure could offer a powerful strategy to overcome some genetic diseases.

Although therapeutic cloning remains theoretical, many scientists are optimistic that the technique will eventually prove useful and believe it holds the most promise for treating ailments where a single cell type is affected. Frequently mentioned conditions include type 1 diabetes, which results when pancreatic islet cells fail to produce sufficient insulin, and Parkinson's disease, which results from the death or impairment of a specific class of neurons. It is less clear how the therapeutic cloning procedure could directly benefit patients with more complex ailments affecting many cell types, such as Alzheimer's disease. This is not to say that scientists aren't excited about using somatic cell nuclear transfer technology to study Alzheimer's and other complex diseases: they are. To take Alzheimer's as an example, scientists would like to compare the development of neurons from healthy and diseased individuals. Examining the differences in neural development may yield insights into how Alzheimer's develops. To make this comparison, scientists would need to create a cloned embryo from a patient with Alzheimer's disease and use it to derive a human embryonic stem cell line.

Although nuclear transfer has not yet been used to create disease-specific embryonic stem cell lines, scientists in both the United States and United Kingdom are working toward this goal and aiming to derive embryonic stem cell lines from patients with diabetes and amyotrophic lateral sclerosis (or motor neuron disease). Scientists have developed disease-specific embryonic stem cell lines for a few genetic illnesses. These lines were not created by cloning but by using embryos created for *in vitro* fertilization but discarded after genetic screening indicated they contained defects. A group of scientists, led by Yury Verlinksy, at the Reproductive Genetics Institute in Chicago created eighteen human embryonic stem cell lines from embryos affected with a variety of genetic disorders including Duchenne muscular dystrophy, fragile-X syndrome, and Huntington's disease.[4]

Proof in principle (in mice)

Although therapeutic cloning in humans remains theoretical, scientists have made significant progress in animals. This is crucial, because animal research is an important step in the development of therapies for human use. Before a new therapy is approved for use in humans, regulators want evidence that it is safe and beneficial in animals.

Most of the animal research related to potential human embryonic stem cell based therapies has been in mice. A number of research groups are active in the field and refining various elements of the technique, and one group has gone further and shown that the entire therapeutic cloning procedure can work in mice.[5] This research was led by Rudolf Jaenisch and George Daley at the Whitehead Institute for Biomedical Research, in Cambridge, Massachusetts, just outside Boston. Although this proof in principle research took place in mice, scientists know that what can be done in mice can almost always eventually be done in humans. The successful use of therapeutic cloning in mice bodes well for its use in humans in the future.

You may recall that one of the advantages of working with mice is that scientists have, literally, thousands of different strains from which to choose. In this research, the scientists started with a particular strain of mice that exhibited a well-defined genetic disorder – severe immunodeficiency – to test the therapeutic cloning protocol. This particular immune deficiency results because the mice do not express the gene Rag2, but the immune system of Rag2-deficient mice can be rescued by bone marrow transplant from genetically identical healthy donors. This ability makes Rag2-deficient mice an ideal system to test the therapeutic cloning procedure. This particular genetic disorder resembles the human "Omenn syndrome," a rare but serious genetic illness.

Cloning the immunodeficient mice was the first step in this research. To do this, the scientists isolated cells from the tip of a mouse's tail and used them as nuclear donors for transfer into enucleated eggs. Twenty-seven of 202 reconstructed eggs developed successfully to the blastocyst stage, and the scientists successfully isolated one mouse embryonic stem cell line from these blastocysts. Next, the scientists repaired the Rag2 gene *in vitro*, using standard genetic engineering techniques to introduce a fully functional version of the gene into the cells. They inserted a selectable marker gene together with the functional Rag2 gene, which allowed the selection of cells that had been successfully modified. This gave them a population of mouse embryonic stem cells in which the original genetic defect had been fixed. The scientists verified that these cells were fully repaired by generating new mice using the tetraploid embryo complementation technique. The mice derived from the genetically repaired embryonic stem cells had fully functional immune systems. Of course, the idea of therapeutic cloning in humans is to cure living individuals, not to create healthy clones of sick people, so Jaenisch and his colleagues went on to show that these repaired embryonic stem cells could be used in a transplant therapy.

A key challenge associated with any therapeutic use of embryonic stem cells is controlling their differentiation. The power of embryonic stem cells comes from their ability to form any cell type but it is a power that poses difficulties. In particular, the differentiation of embryonic stem cells into mature cells has proven difficult to control precisely. In this case, however, the scientists had devised a means to convert mouse embryonic stem cells preferentially, although not exclusively, into hematopoietic stem cells which, under appropriate conditions, can restore the immune system of immune-deficient mice. To test the therapeutic cloning protocol further, the scientists derived these hematopoietic stem cells from their genetically repaired mouse embryonic stem cell line.

To conclude their experiment, they attempted to transplant these cells into the original mutant mice. The scientists encountered more difficulties with this step than they had anticipated.

Their initial transplants met with little success, and despite being genetically identical to the recipient organism, were rejected by the recipient's immune system. Jaenisch and his colleagues believe this rejection was due to specific characteristics of the Rag2 mutant, rather than any inherent properties of the nuclear-transfer derived embryonic stem cell line. To bolster this claim, they repeated the transplants after depleting the particular cell type that they believe caused the rejection and reported a limited restoration of immune system function. They went further, transplanting their genetically repaired cells into Rag2 mutant mice engineered to lack the cells they believed were causing the rejection, and the repaired embryonic stem cells successfully restored most but not all, of the immune system's functionality.

Despite the difficulties encountered during the transplant stage, this experiment illustrates the potential power of therapeutic cloning. Although the procedure has many steps, each of which is challenging, this research suggests that the entire protocol can be accomplished and offers hope that with time, it may be refined and used in humans.

Baby steps in the battle against Parkinson's disease

Parkinson's disease is among the diseases most frequently mentioned as a target for human embryonic stem cell based therapies. This optimism is because Parkinson's symptoms are caused by the loss of a single cell type – midbrain dopamine neurons. Although embryonic stem cell based treatments of this condition remain in the distant future, scientists have made important progress in animals. A group of scientists, led by Ron McKay, a Scot now working at the U.S. National Institutes of Health, has shown that mouse embryonic stem cells can be used to improve the condition of mice with an ailment similar to

Parkinson's disease.[6] They used a five-step protocol to induce genetically modified mouse embryonic stem cells to differentiate preferentially into dopamine neurons. These neurons were then transplanted into mice that lacked dopamine neurons in a key section of their brain and so exhibited Parkinson's-like symptoms on one side of their bodies.

To test the effectiveness of this therapy, the scientists compared the mobility of the animals receiving transplants with a control group of mice receiving "sham" operations. (To distinguish cause and effect in scientific experiments, it is crucial to use an appropriate control. The use of animals that received "sham" transplants of unrelated cells provides verification that any changes were due to the transplanted neurons and not to the regeneration that occurs naturally following an invasive operation.) Nine weeks after the transplant, the scientists found significant differences between the two groups of mice. Mice that had received transplants of dopamine neurons performed markedly better on a series of tests measuring mobility and the use of the paw on the animal's damaged side. Although these mice were certainly not cured, their improvement was significant and promising. The research showed conclusively that dopamine neurons derived from embryonic stem cells could, at least in mice, survive transplant and lead to partial improvement of Parkinson's-like symptoms. Further research is necessary, in both mice and non-human primates, to verify the long-term safety and effectiveness of this sort of therapy.

Successes like the two discussed above raise hopes that human embryonic stem cells will one day be used therapeutically. However, despite these and other successes in mice it is important to recognize that translating this research to work with humans is neither simple nor quick. Scientists have worked with mouse embryonic stem cells for nearly twenty years more than they have worked with the equivalent human cells and it is only natural that knowledge of mouse cells is significantly more advanced. Human research is also limited by the controversy it engenders. While many scientists are optimistic about human therapeutic cloning and embryonic stem cell based therapies in the long term, few believe any therapies are imminent.

The South Korean saga

A cautionary tale regarding the hype and hope surrounding human embryonic stem cell research is provided by the events that unfolded recently in South Korea. In two years, South Korean scientists went from fringe players in this nascent field to unquestioned world leaders. Then, in weeks, their success story unraveled amid allegations of ethical improprieties and outright fraud. The story centers on the laboratory run by Woo Suk Hwang at Seoul National University but because at its high point Hwang's success came to symbolize the growing stature of South Korean science, its implications extend far beyond.

Hwang, who trained as a veterinary scientist, built his reputation cloning cows and pigs in the late 1990s. Because of his stellar reputation, his first reported foray into human therapeutic cloning was greeted with modest surprise, but not shock. In a February 2004 paper in *Science*, a top peer-reviewed journal, Hwang and his collaborators announced that they had cloned thirty human embryos and successfully derived a human embryonic stem cell line from one of them. This report apparently marked the first time a human embryonic stem cell line had been derived from a cloned embryo and was widely viewed as a key step in the development of therapeutic cloning. In their paper, Hwang and his co-authors indicated that they had used 242 human eggs from sixteen uncompensated donors. Scientists were impressed with the number of human eggs available for this research, as limited access to eggs had hindered research elsewhere. However, the inefficiency of the process – nearly 250 eggs were required to create a single cell line – worried observers and called into question the economics of therapeutic cloning.

Following publication of this landmark paper, 2004 was a year of increasing worldwide fame for Hwang. The only inkling of trouble was a minor controversy over some of the eggs used in the research. The controversy started when a Ph.D. student in Hwang's lab told *Nature*, another leading science journal, that she and another woman in the lab had donated eggs for the project.[7] Bioethicists consider egg donation by students or other junior researchers to be ethically inappropriate, because of the possibility of coercion inherent in such an arrangement. The student later retracted her statement, blaming her poor English for the misunderstanding. Some questioned this retraction, however, since the student had, in the initial interview, named the hospital where the donation occurred and clearly explained her rationale for donating. Despite calls for an investigation, these concerns did little to slow Hwang's growing momentum.

Hwang's lab continued to publish dramatic advances in 2005. First, in May, they published a second therapeutic cloning paper in *Science*, detailing rapid progress made since the previous year. In this paper, they reported both the derivation of eleven new patient-specific human embryonic stem cell lines from cloned embryos and a huge increase in the efficiency of the therapeutic cloning process. Deriving a new patient-matched human embryonic stem cell line now required an average of only twelve to seventeen eggs, rather than nearly 250. As well as apparently validating the original paper, this increase in efficiency was seen as a crucial advancement, moving therapeutic cloning toward economic feasibility much more quickly than anyone had expected. Although twenty-four of the twenty-five researchers on the team reporting these advances were based in South Korea, Gerald Schatten, the University of Pittsburgh developmental biologist who only two years previously had downplayed the possibility of creating embryonic stem cell lines from cloned primate embryos, was prominently listed as the paper's senior author. Following publication of this paper, Hwang's stature continued to grow. In August 2005, the birth of Snuppy, the world's first cloned dog, was announced, further cementing Hwang's position as the world's leading cloning expert. In mid-October, the South Korean government launched the World Stem Cell Hub to distribute human embryonic stem cells around the world and help scientists derive new human embryonic stem cell lines from cloned embryos. Not surprisingly, Hwang was appointed as the network's first head.

Hwang's leading role in the scientific world unraveled rapidly in late 2005. In early November, Schatten alerted *Science* to reports in the South Korean media that one of Hwang's collaborators had paid for eggs for use in their research. Then on November 12, he dramatically cut ties with Hwang, citing new information suggesting that "misrepresentations might have occurred relating to [egg] donation" as a reason to break his and Hwang's 20-month long collaboration.[8] Schatten's decision gave new traction to allegations against Hwang. Spurred on by reports in the South Korean media, two weeks later Hwang admitted that his research had relied on eggs from paid donors and junior members of his team. He claimed to have been unaware of these ethical breaches when they occurred and apologized for not coming clean once he was informed.

The ethical shortcomings, while serious, did not affect the scientific conclusions of the papers and scientists continued to believe Hwang's reports represented the state of the art in therapeutic cloning. However, in December, allegations of outright fraud began to surface. A Korean TV network reported that at least one of Hwang's patient-specific embryonic stem cell lines didn't genetically match the tissue sample from which it was supposedly derived. A few days later, Hwang notified *Science* that some of the pictures published as part of the 2005 paper were incorrect. These events convinced the University of Pittsburgh and Seoul National University to open investigations into the research. Problems with the papers appeared with increasing frequency. On December 13, Schatten asked Hwang to retract the May 2005 paper. In a letter written to *Science*, Schatten said, "My careful re-evaluations of published figures and tables, along with new problematic information, now casts substantial doubts about the paper's accuracy."[9] Three days later, Hwang held a press briefing where he acknowledged that mistakes were made and said he would ask *Science* to retract the paper. He maintained, however, that his lab had developed the technology to create human embryonic stem cell lines from cloned embryos.

A month later, the investigative committee at Seoul National University delivered its report. Both of Hwang's landmark stem cell papers were declared deliberate frauds. The committee concluded that Hwang's team had never successfully created a human embryonic stem cell line from a cloned embryo and had manipulated data in both the 2004 and 2005 publications to make it appear that they had. The report suggested the one cell line reported in the 2004 paper was more likely to have been derived from parthenogenesis, a form of asexual reproduction, than cloning. They did conclude that Hwang's team possessed the ability to grow cloned human embryos to the blastocyst stage, an important advance in its own right. They also, as has already been mentioned, verified that Snuppy was indeed a clone.

While this report was an embarrassment to the South Korean government, which had strongly supported Woo Suk Hwang's research, its long-term impact on biomedical research in the country remains to be seen. Regardless, it marked the end of South Korea's rapid move to the forefront of therapeutic cloning technology.

Hurdles to human therapeutic cloning

The prospects for human therapeutic cloning, in the short term, took a significant turn for the worse when the South Korean research was deemed fraudulent. Just as many scientists were starting to believe that a key challenge involved in therapeutic cloning – the efficient creation of patient-matched embryonic stem cells – was feasible, it became clear that human embryonic stem cell lines had never been derived from cloned human embryos. Scientists were left with only a few reports of cloned embryos grown successfully to the blastocyst stage. They now must repeat the work everyone believed Hwang had successfully completed.

Leaders in the field include Alison Murdoch at the University of Newcastle upon Tyne in the United Kingdom, Miodrag Stojkovic at the Prince Felipe Research Center in Spain, Guangxiu Lu of the Xiangya Medical College in China, and scientists working at Advanced Cell Technology, a U.S. based biotechnology company. Each of these groups has reported progress creating cloned human embryos but none has successfully isolated embryonic stem cells from them. Other scientists, who had chosen to study different aspects of human embryonic stem cell science, are refocusing their research efforts on the therapeutic cloning protocol. As of April 2006, seven groups, three in the United States, three in Europe, and one in China had announced plans to attempt to create human embryonic stem cell lines from cloned embryos.[10]

Assuming these hurdles can be overcome, another big challenge remains: controlling the differentiation of human embryonic stem cells into the appropriate mature cell types for transplantation. Progress has been made; scientists have observed cells from a wide range of therapeutically interesting tissues among cell populations derived from human embryonic stem cells and begun to explore how to control this differentiation. However, in most cases, protocols to yield pure or relatively pure populations of particular cell types do not yet exist or need refinement. In McKay's mouse experiment, for instance, scientists used a five-step procedure to yield a cell population significantly enriched in a particular type of neuron. Such procedures will need to be developed for essentially any cell type that scientists want to transplant. Since developing a single reliable differentiation proto-col can take several years, this is a significant challenge.

The magnitude of the remaining challenges should not leave you with the impression that scientists have made little progress toward the development of therapies. In addition to research in mice and other animals, scientists have made important strides developing culture systems for human embryonic stem cells. When James Thomson first grew human embryonic stem cell lines in a petri dish, he placed the cells on a layer of mouse cells. While this approach is effective, it risks contaminating the human embryonic stem cells with animal products. The potential transfer of animal-specific infectious agents, such as viruses, across species boundaries greatly complicates the clinical use of these human cells. In the United States, therapeutic substances that have been in contact with non-human animal products must pass the Food and Drug Administration's special xenotransplanta-tion guidelines before approval. Although the initial human embryonic stem cell lines were derived in the presence of mouse cells, scientists have now learned to create feeder layers of human cells. This advance, reported by a number of research groups, including those led by Bongso and Thomson, represents an important step toward eventual therapeutic use of human embry-onic stem cells, including, perhaps, those created by therapeutic cloning.

Further reading

A thorough and easy-to-read review of the history and promise of stem cell research can be found in Ann Parson's *The Proteus Effect: Stem Cells and Their Promise for Medicine* (Joseph Henry Press, 2004). Parson chronicles the early history of stem cell research, focusing on not just embryonic stem cells but also important adult stem cells that have been studied for many years and play key roles in bone marrow transplantation and other therapies.

For those more interested in a detailed but still accessible treatment of the science of human embryonic stem cells, Ann Kiessling, a professor at Harvard Medical School, and Scott Anderson, a science writer, have collaborated on *Human Embryonic Stem Cells: An Introduction to the Science and Therapeutic Potential* (Jones and Bartlett Publishers, 2003). This book is a more challenging read than Parson's *Proteus Effect* but will reward dedicated readers with an excellent overview of the science. A newly revised second edition was published in late 2006.

Another option to consider is Christopher Thomas Scott's *Stem Cell Now: From the Experiment that Shook the World to the New Politics of Life* (Pi Press, 2005). Scott offers an introduction to stem cell science and some of the ethical arguments in the debate over stem cell policy.

6

The ethical debate over human cloning

Despite the benefits cloning technology offers society – both through animal cloning today and perhaps cloning for medical research in the future – the debate over the technology has been dominated by the possibility that a cloned human being may one day be born. No such person has yet been born; at least, no such birth has been acknowledged and confirmed. However, there is little doubt that continuing advances in cloning technology make future attempts to clone humans increasingly likely to succeed.

There is an almost universal consensus among mainstream scientists that cloning humans for reproductive purposes is too dangerous to attempt at the current time. Thus, there is little ethical debate over human reproductive cloning today. Less agreement exists, however, on the question of whether human reproductive cloning would be ethically acceptable, assuming the technology was refined to the point where it was as safe as or safer than traditional reproduction. Nor is there a consensus on the ethical acceptability of cloning to create embryos for medical research (the therapeutic cloning protocol discussed in Chapter 5). This chapter will review some of the arguments both for and against the potential cloning of humans, as well as the major events that have shaped this debate.

Cloning claims

For better or worse, public perception of human reproductive cloning and debates over the technology have been in part defined by those who have expressed interest in the technology or claimed to have produced cloned humans. Most notable are the Raelians. This religious sect was founded by Claude Vorilhon – renamed Rael – in 1973. The sect's founding followed what Rael describes as a meeting with four-foot-tall, green-skinned, extra-terrestrials on a French mountaintop. The Raelians announced the creation of Clonaid, a company dedicated to cloning human beings, shortly after Dolly's birth. Raelians, who believe aliens created humans in a laboratory, view cloning as a key to eternal life.

Clonaid brought worldwide attention to the Raelians in 2002 when it announced that its research had led to the birth of the world's first cloned human child, a baby girl named Eve. Brigitte Boisselier, Clonaid's chief scientist, made this dramatic announcement not in a peer-reviewed journal but at a press conference in a Florida hotel. The news, along with her claim that four other women were pregnant with cloned human embryos, was greeted with rampant skepticism. This skepticism was partly held in check by promises that the baby and her mother would be subject to independent DNA fingerprint testing within two weeks. Because these tests promised to show clearly if Eve was a clone of her 31-year-old American mother, as Clonaid claimed, and because some scientists suggested human cloning might well be possible if a group was willing to tolerate a high failure rate, the Raelians' claims of success garnered significant media attention.

In retrospect, the press conference's location in Hollywood, Florida seems oddly appropriate, as many believe the announcement was more show business than science. No evidence to prove Eve was a clone – or that she even existed – was ever produced. A well-regarded science journalist, who initially agreed to oversee genetic tests to assess the group's claims, abruptly abandoned them, suggesting the project might be an "elaborate hoax."[1] Questions have also arisen about Clonaid's research facilities: although the Raelians claim to have extensive financial resources and to have raised some seven million dollars toward construction of an embassy to welcome humanity's extra-terrestrial creators back to earth, no evidence exists that they operate any laboratory facilities suitable for an attempt to clone humans.

Regardless of any scientific accomplishments, Clonaid proved a remarkable public relations success for the Raelians. Rael, in his self-published book, *yes to human cloning*, reports that for an investment of three thousand dollars his group received some fifteen million dollars worth of media coverage and the opportunities to testify before both the U.S. Congress and the U.S. National Academy of Sciences. Although Clonaid and the Raelians have largely fallen into obscurity as fears of imminent human cloning have receded somewhat in recent years, the group continues to court cloning-related publicity. In early 2006, for instance, the group reportedly offered the discredited South Korean cloning pioneer Woo Suk Hwang a job.

Among the others who have risen to prominence by claiming interest in cloning human beings is Richard Seed, a Harvard-educated physicist. Seed, who first announced his plans during the question and answer session of a conference on the legal and ethical issues surrounding human cloning in December 1997, gained worldwide attention when his plan to open what he called the Human Clone Clinic was featured on National Public Radio in the United States in January 1998.[2] Seed, who had previously attempted to enter the fertility business in the 1980s, indicated he was seeking venture capital for his clinic and aimed initially to serve some five to ten thousand infertile couples in the United States who could not conceive using existing methods, including *in vitro* fertilization. Although numerous scientists denounced Seed's efforts and suggested he was ill-qualified to proceed with the project, he claimed a moral imperative for continuing, stating in his radio interview that "cloning and the reprogramming of DNA is the first serious step in becoming one with God."

Following his brief stint in the limelight, Seed and his cloning plans have faded from public attention, presumably because he had neither the funding nor the expertise necessary to put his plan into action. Rael reportedly offered to bankroll Seed's project in early 1998, although it is not clear if the two cloning mavericks entered into any serious discussions.

The final significant human cloning claim came from a more orthodox and believable source. In January 2001, Panos Zavos, who has held positions in reproductive physiology at the University of Kentucky and co-founded a fertility clinic in Lexington, Kentucky, announced that he was collaborating with Severino Antinori, an Italian fertility doctor, to produce a cloned human baby. Zavos stated that human cloning was inevitable and argued that it was preferable for the technology to be developed by professionals as his rationale for proceeding. Zavos and Antinori's scientific expertise – both had published numerous scientific papers in related fields – lent credibility to the notion that human cloning was imminent, while their reputations as renegades suggested they might be willing to buck public opinion and medical consensus to accomplish their goals. Antinori, for example, who first gained international notoriety in 1994 when he used *in vitro* fertilization to help a sixty-three-year-old woman become pregnant and give birth to a child, was no stranger to controversial science.

Although their collaboration faltered, both Zavos and Antinori stated they were continuing efforts to produce a cloned human child and at various times announced progress toward this goal. In April 2002, media reports surfaced that a woman Antinori had been treating was eight weeks pregnant with a cloned human embryo. These reports went unconfirmed and no mention was made of a child resulting from this pregnancy. Two years later, in January 2004, Zavos reported transferring a cloned human embryo into the womb of a thirty-five-year-old woman but announced a few weeks later that the attempt had been unsuccessful. Months later, Antinori again claimed the headlines, when he announced that he knew at least three cloned human babies had been born but refused to provide proof or any additional details to support these claims.

More recently, Zavos has published a detailed description of his 2004 attempts to produce a cloned human child to help an infertile couple reproduce. Zavos reported using cultured skin cells from the infertile male as donors for nuclear transfer into enucleated eggs retrieved from the man's wife.[3] One of three reconstructed eggs developed to the four-cell stage and was transferred to the women's uterus. This cloned embryo did not implant and no pregnancy resulted.

The chance remains that a cloned human being has been born and is living in unannounced anonymity. However, the lack of proof associated with these claims, along with the clear desire for publicity shown by the various players, suggests none of these efforts have proved successful. They remain important, due to their role in shaping the public perception of cloning. Clonaid, the company created by the Raelians, has been mentioned nearly fifty times in magazines, such as *Newsweek, U.S. News & World Report*, and *The Economist*, since 2000. Antinori likewise continues to court publicity and each media appearance, such as coverage of his role helping a sixty-three-year-old British woman become pregnant in early 2006, brings his cloning claims back to public attention.

These claims attract attention in part because many scientists believe human cloning is essentially a numbers game.[4] A well-financed and dedicated group of scientists might well succeed if they could recruit enough women willing to donate their eggs and wombs to the effort. Such a group would need to persevere through the many failures that would almost inevitably occur. Although the cloning claims made by the Raelians remain unconfirmed and are widely believed to be fraudulent, this sort of group, with numerous followers willing to "volunteer" for such a project at their leader's suggestion, and with the funding to support a few maverick scientists in search of notoriety, may well produce the world's first cloned human being.

The consensus against human cloning (for now)

With these few exceptions and no doubt some others who choose not to share their thoughts with the world, scientists, bioethicists, and policymakers are nearly all in agreement that human reproductive cloning should not currently be attempted. This consensus derives from concerns about the safety of cloning and the risks the technique might pose to the fetus and developing child. If experiences with animals are extrapolated directly to human cloning, it seems likely that many cloned embryos would develop abnormally and spontaneously abort. (Of course, this also occurs in normal sexual reproduction; an estimated 50 percent of developing embryos fail to implant and many more spontaneously abort following implantation.) More concerning, to many observers, is the risk that some substantial portion of cloned embryos might develop to term but be born with potentially serious developmental abnormalities. Approximately three percent of children born in the United States and other developed countries suffer from serious birth defects, but this rate would likely be much higher for children produced asexually through cloning technology.

The extent to which advances in animal cloning will reduce these safety concerns remains an open question. Some believe that cloning is inherently unsafe and that these concerns will persist indefinitely. The President's Council on Bioethics, an advisory committee charged with providing bioethics advice to President Bush, concluded that, because human reproductive cloning is unsafe today and because experiments to make it less safe were also unethical, "there seems to be no ethical way to try to discover whether cloning to produce children can become safe, now or in the future."[5]

Many scientists reject this view and are willing to consider the possibility that experiments on animals could advance cloning technology to the point where the technique was safe enough for human use. This view squares more fully with the history of exponential advance in molecular biology. The rapid and frequent advances made in the fifty years since DNA's structure was discovered lend credence to the oft-repeated claim that, when looking to the future, scientists and other prognosticators tend to overestimate the short-term impact of technological change but underestimate the long-term impact. Before genetic engineering was first demonstrated or the human genome was sequenced, noted scientists said these feats were not just difficult but impossible yet these, and similar advances, were achieved in relatively short order. This suggests that those who argue cloning will never be safe enough for human use should tread cautiously.

Given the uncertain future of cloning technology and its potential application to human reproduction, significant attention has been paid to the question of whether reproductive cloning should be allowed, if it were safe for human use. In this, consensus is the exception rather than the rule. Almost everyone agrees that the cloning of human beings raises novel and interesting ethical questions but when balancing the potential reasons to permit reproductive cloning against the potential reasons to prohibit it, different people come to different conclusions.

Cloning scenarios

To examine this debate, we will look briefly at several potential situations in which some believe cloning might be appropriate. These scenarios vary widely and many advocates of human reproductive cloning support the technology in some, but not all, of them.

One potential use is as a tool to help combat infertility. Human reproductive cloning is often seen as an extension to the existing arsenal of assisted reproductive technologies. In many cases, *in vitro* fertilization and related techniques allow parents to have a child who is genetically related to them both, but for some types of infertility such a solution is not possible. Consider a couple where the husband, due to complications associated with testicular cancer, produces no sperm and is thus not even a candidate for variants of *in vitro* fertilization that work with a single sperm cell. The standard option for this couple would be donated sperm but this means the child would not be genetically related to his or her father. This couple might prefer to use cloning: the wife could provide eggs and the infertile husband could provide somatic donor cells. The resulting child, though a "delayed" identical twin of his father, would contain his mother's mtDNA, and thus represent, in some senses, a combination of his parents. This hypothetical couple is similar to the one Panos Zavos reported trying to help conceive through cloning in 2004.

Another potential use would be to replace a lost child. Supporters of this scenario suggest that cloning a child who died early in life might help parents overcome their tragic loss. It would probably be easier for such parents simply to have another child using traditional methods but the parents might prefer to use cloning, for one reason or another. It might even be the case that this couple could no longer reproduce: perhaps the mother's ovaries had been removed to counteract a genetic predisposition to ovarian and breast cancer. Cloning may be the only means for such a couple to have another child genetically related to both parents.

Others have argued that reproductive cloning may be acceptable when a couple wants to have another child who can serve as a tissue donor for an older child with a serious but treatable illness. Hypothetically, a couple could be planning to have another child when they find out their existing son has a rare acquired blood disorder that can only be treated by a bone marrow transplant from a genetically matched donor. There is a slight chance a naturally conceived child would be a genetic match but cloning their son would guarantee it. In this couple's mind, cloning is the best option. They both give their son a younger brother and save his life.

Reproductive cloning could also potentially provide a means for people not in traditional heterosexual relationships to have genetically related children. Some have suggested that a single woman might reproduce by cloning herself rather than relying on donor sperm; others have argued that reproductive cloning might provide a feasible means for same-sex couples to reproduce. This is particularly true for lesbian couples, who could divide up the egg donor, somatic cell donor and surrogate mother roles. Homosexual male couples would need to find both egg donors and surrogates.

Arguments for human cloning

The exceptional scenarios described above represent a sampling of situations in which some people believe human reproductive cloning might be ethically acceptable. Not everyone agrees about these scenarios. Some support cloning in one or two cases, while others reject it for all these situations. Beyond these (and other) specific cases, a number of general principles, including personal liberty, reproductive choice, and open scientific inquiry, support the notion that human reproductive cloning should not be prohibited. For the most part, these ideas don't explicitly encourage human reproductive cloning but argue that, for one reason or another, it shouldn't be banned.

Personal liberty refers to the general preference of many democratic societies that few restrictions should be imposed by the government or other authorities. Since different people have different preferences, maintaining as wide a sphere as possible in which individuals are free to select their own actions is seen as serving the greatest good. Such an argument has limits. Few societies condone murder, even if an individual deems it serves his or his personal interest. However, human reproductive cloning is not murder and some supporters argue that, in the absence of evidence of significant harm, decisions regarding its use should be left to individuals, not the government.

Some supporters of human reproductive cloning argue that because it is a form of human reproduction, it falls in a special range of activities that must be actively protected from government interference. This idea of "reproductive freedom" is particularly entrenched in the United States, where the Supreme Court has written: "If the right of privacy means anything, it is the right of the individual, married or single, to be free from unwarranted governmental intrusion into matters so affecting a person as a decision whether to bear or beget a child."[6] This ruling suggests to some legal scholars that a ban on human reproductive cloning in the United States, as has been proposed and debated from time to time, would not be constitutional. For some couples, cloning may provide the only means for them to have a genetically related child. Banning human cloning could be construed as an unwarranted intrusion into their personal decision-making process. The court decision defining this right to reproductive freedom, written before *in vitro* fertilization existed, understandably omitted any mention of the specific method a person used to bear or beget a child, but it can be plausibly argued that a right to reproductive freedom should cover not just traditional sexual reproduction but any means a person might choose.

Scientific inquiry is viewed as a public good and some have suggested that unrestricted scientific inquiry should be permitted and encouraged to the greatest possible extent. If research related to human reproductive cloning represents a legitimate field of scientific inquiry, this principle suggests it should be allowed to proceed. This idea derives from the observation that relatively unfettered scientific investigation has had tremendous, and largely positive, impacts on society over the last few centuries. Given this history, and in the absence of compelling reasons to limit a line of research, allowing scientific inquiry to proceed without restrictions best serves the public's interest. However, while scientific inquiry proceeds with relatively few limits in many societies, it is rarely completely unrestricted. Indeed, following a series of ethically inappropriate research projects, including the infamous Tuskegee Syphilis Study in the United States and published exposés of questionable research practices in both the United States and the United Kingdom, scientific oversight designed to protect research subjects is now the norm in most countries.

Arguments against human cloning

A number of arguments against the use of cloning technology are lined up against the would-be parents and the general principles supporting human cloning, but because humans have not yet been cloned and because cloned animals can't tell us how they feel, they are largely hypothetical. That is, they are based on bioethicists' guesses of how cloned human beings would feel or the impact cloning might have. These arguments address a number of issues, including concerns about the individuality of cloned humans, the impact of cloning on families, and the potential for cloning to encourage the objectification of people and lead toward a new era of eugenics.

Some suggest that cloned human beings would suffer from the lack of a unique genetic identity. These critics worry that cloned humans would be constantly compared to their genetic progenitors and suffer from unrealistic expectations created by these comparisons. In essence, these cloned human beings might feel as if their lives had already been lived. Cloning could violate what some ethicists have called a "right to an open future." Such concerns do not apply to identical twins because, while they share a genetic identity, their lives unfold at the same time, giving each the freedom to escape from the other's shadow and live their own life. In cloning, the older twin's life would have already unfolded, and the latter-born twin could never escape this shadow. Critics of this argument point out that it relies on a rather crude genetic determinism. Genes are important but they do not define our existence. Studies of identical twins consistently find that while they share many characteristics, they also differ in important ways. If this is the case for identical twins who shared the same womb and grew up in the same household, it seems only fair to guess that a latter-born twin, who developed in a different uterine environment and grew up years later, would differ significantly from his or her genetic donor.

Critics of human reproductive cloning also worry about the impact of cloning on family structure. If an infertile couple chose to have a son by cloning the father, the family would be, in genetic terms, quite unusual. Genetically, father and son would be twin brothers and the father's parents – ostensibly the cloned child's grandparents – would truly be his genetic parents. The mother, though she gave birth to her son, would have few genetic links with him. Just what impact these atypical circumstances might have is not clear but some have suggested that the cloned child's close genetic ties to only one parent might complicate family dynamics and perhaps reduce the stability of the family. In other cases, the impact on the family might be less pronounced. Cloning an existing child, to replace a lost child or create a compatible tissue donor, is less problematic for family structure; the cloned child is a younger sibling of the donor and the standard parent-child relationship remains intact.

These concerns are countered, to some degree, by the observation that humans are quite flexible in their family structures. High-tech reproductive technologies, such as *in vitro* fertilization using donated eggs, surrogacy, and circumstances such as divorce or adoption lead to structures that differ from the traditional nuclear family. If asked, most parents would no doubt prefer a traditional structure but this isn't always possible. Healthy children and strong families have resulted from all the variations listed here, and countless others.

Because children born through cloning would have their entire genome selected for them by their parents, some critics believe the technology opens the door to the objectification of human beings. Rather than loving their children as gifts and discovering their potential as they grow and develop, parents may begin to view them as products that can essentially be made to order. These critics fear that human cloning moves society toward a situation where children are manufactured rather than begotten. Like other manufactured products, these humans, produced through the transfer of a somatic cell nucleus into an enucleated egg, might be treated with less respect than humans produced through the fusion of a sperm and egg. Even if this tendency toward objectification is not strong at first, cloning might worsen it. Human reproductive cloning could open the door to the genetic engineering of humans (remember that the scientists who cloned Dolly were looking for an efficient way to produce genetically engineered cows) and genetic engineering could greatly exacerbate this objectification. Parents who chose specific genes for a child might well have high expectations and express disappointment with the manufacturing process – or even the child – if results weren't as expected.

Some believe these fears are overstated. Genetic screening is relatively common and nobody suggests that a child born following pre-implantation genetic diagnosis (used to screen embryos for certain genetic characteristics after *in vitro* fertilization) is any less human or less deserving of respect. Couples would probably turn to cloning as a last resort and be tremendously grateful that a child was born at all, not upset that he or she didn't meet their expectations. Even some supporters of human reproductive cloning agree that cloned children might face more specific expectations than children produced the old-fashioned way but they claim these expectations are neither new nor cause for particular concern. Every parent has dreams and hopes for their children and many spend thousands of dollars on private schooling, music lessons, and the like to help their children fulfill these dreams but few, if any, parents love their children any less when a particular hope goes unfulfilled. It is not clear why parents of cloned children would be any different.

Beyond these concerns, which apply to even a single cloned human being, some argue that if used more broadly, cloning could lead the way to a new era of eugenics. Eugenics, simply defined, is the application of selective breeding to human beings in the hope of improving the human race. Although eugenic ideas have a history dating back to the ancient Greeks, the movement came to prominence in the West in the late nineteenth century, when Francis Galton, a cousin of Charles Darwin, tried to apply the theory of evolution by natural selection to human breeding. In the United States, eugenic policies, based on a poor understanding of human genetics, were common in the early twentieth century and led to the compulsory sterilization of some 60,000 people. Eugenic thinking, taken to its horrific extreme, led to the Holocaust in Nazi Germany during World War II.

These eugenic fears are stoked by misconceptions about cloning science. Despite Hollywood's best efforts, armies of identical cloned soldiers are unlikely to be training at secret military bases anytime soon, nor will factories be operated by hundreds of specially bred genetically identical brothers. Misconceptions about governments also play a role. Eugenics movements relied on governmental coercion and it might not even make sense to speak of eugenics in an era where, at least in developed Western countries, state control of reproduction is exceedingly unlikely.

Those who fear cloning could lead to a new era of eugenics don't worry so much about a return to the government-imposed negative eugenics of the past. Instead, they express concern that eugenic ends could result from individual choices. This argument relies on the use of cloning technology to facilitate the genetic engineering of humans. If such a practice was widespread, genetic traits that were selected for in large numbers of children might become more common in the population over time. Similarly, traits that were selected against might become less common. This could well be the case. However, even if human cloning was widespread, it is not clear that everyone would choose to select for or against the same traits. Parents have different ideas about what is good for their children and they might well use cloning technology to select for different characteristics.

Agreeing to disagree

Beyond the consensus over the health risks associated with human reproductive cloning, it has proved difficult to reconcile the various arguments for and against this still-hypothetical technology. Scientific organizations issuing reports on the subject of human reproductive cloning typically focus on short-term health concerns and ignore, for the most part, the challenging ethical questions raised by assuming future safety and efficiency improvements. Meanwhile, ethical advisory committees have divergent views on this issue.

In the United Kingdom, the Royal Society has called for an international ban on human reproductive cloning, largely on safety grounds, while recognizing that research is continuing, at least on the fringes of the scientific community. To address this, the Society compiled a checklist of statements that outline how human reproductive cloning research should proceed, if it is to continue. This checklist aims to reduce the hype associated with unsubstantiated claims of human cloning, such as those made by the Raelians. In the United States, a comparable group, the U.S. National Academy of Sciences, has published a report on the scientific and medical aspects of reproductive cloning. This group also rejected human reproductive cloning on safety grounds but called for a review within five years, to re-examine the animal data justifying this ban. Beyond calling for a broad dialogue, the report explicitly ignored ethical issues.

The first major ethical report on human cloning was produced by the National Bioethics Advisory Commission in the United States. This diverse group of scientists, medical practitioners, and bioethicists was chaired by Harold Shapiro, an economist and past president of both the University of Michigan and Princeton University. It was formed to advise President Clinton on ethical questions related to the life sciences. Their report, published in June 1997, just a few months after Dolly's birth was announced, called for a moratorium on human reproductive cloning in both the public and private sectors. Their recommendation was based more on the health concerns the commission members believed human cloning posed than the other ethical issues they discussed. In a letter sent to President Clinton with the report, Shapiro indicated that:

> Not surprisingly, we have discovered that the potential ability to clone human beings through the somatic cell nuclear transfer techniques raises a whole host of complex and difficult scientific, religious, legal and ethical issues – both new and old. Indeed, the Commission itself is unable to agree at this time on all the ethical issues that surround the issue of cloning human beings in this manner.[7]

A new bioethics committee, formed following President Bush's election, reached different conclusions. This group, chaired by the conservative bioethicist Leon Kass, examined the arguments for and against human reproductive cloning and concluded that: "the Council is in full agreement that cloning-to-produce-children is not only unsafe but also morally unacceptable, and ought not to be attempted."[8] This apparent consensus hid ethical disagreements among the council members. This disagreement was noted parenthetically in the council's report, which while summarizing potential ethical concerns raised by human reproductive cloning, noted that "Different Council Members gave varying moral weight to these different concerns."[9]

Around the world, numerous ethical advisory bodies have called for prohibitions on human reproductive cloning but their reasons remain varied. In France, the National Consultative Ethics Committee for Health and Life Sciences has rejected human reproductive cloning on both safety and moral grounds. After considering potential reasons for allowing cloning, they conclude:

> There is therefore not a single conceivable variation of reproductive cloning of human beings, be it cloning of an adult or of an embryo, which is safe from an accumulation of intractable objections. For all of these reasons, it can only provoke vehement, categorical, and absolute ethical condemnation.[10]

In contrast, bioethicists in Israel, basing their conclusions in part on Judaic teachings, take a more permissive view of human reproductive cloning, assuming the technology is shown to be safe. Michel Revel, the chairman of the Bioethics Committee of the Israel Academy of Sciences and Humanities, has written that, under the theoretical assumption of safety and efficiency, reproductive cloning may be permissible in cases where it provides a clear medical application, such as overcoming otherwise intractable infertility.[11] Reproductive cloning would not generally be acceptable for non-therapeutic uses, such as attempting to re-create someone who had died or assisting same-sex couples to reproduce.

This sampling of some of the conclusions reached by national advisory committees considering human cloning illustrates the complexity of the issues. If expert committees cannot reach a consensus on the ethical appropriateness of reproductive cloning, assuming it is reasonably safe, it is hard to know how societies should resolve these questions. Luckily, safety concerns provide time for deliberation and debate. While scientists proceed with cloning research using animals, individuals and government bodies around the world can continue to debate the ethics of reproductive cloning in the hope of reaching some sort of consensus before the first confirmed cloned human being is born.

Today's world, in which different groups reach vastly different answers to these questions and where those in search of advanced assisted reproductive technologies can freely cross national borders, poses numerous challenges to the regulation of human reproductive cloning. It is even possible that a cloned human being will be born either before a consensus on the technology's use is reached or in some locale where its use is unregulated, and that public policy will evolve in response to this birth, much as it did after the birth of the world's first test-tube baby in 1978.

The ethics of therapeutic cloning

The ethical debate over therapeutic cloning, and human embryonic stem cell research more generally, is less complex but no less contentious than the debate over reproductive cloning. Scientists studying human embryonic stem cells and therapeutic cloning have a noble goal, the alleviation of human suffering. It is not the ends of human embryonic stem cell research but the means that generate disagreement and debate. As we have seen, to move toward this noble goal scientists use pre-implantation human embryos in their research. Although the embryos are donated explicitly for this purpose, if (against the donor's wishes) these embryos were transferred to a uterus, they might survive and develop into healthy children. This possibility, however remote, leads to the ethical question that frames the field: should embryos with some chance of life be used as a means to try to reduce the suffering of others?

This, as in most ethical debates, is a question about which reasonable people can disagree. At its heart, this debate is about differing views of what it means to be a person and whether human embryos deserve full moral status. We grant moral status to an individual or a class of individuals when we acknowledge that their wishes, desires, and rights should be considered in our decision-making. Almost everyone grants full moral status to a healthy child: nobody argues that it is appropriate to harm such a child for our own gain but few grant any moral status to a single human skin cell or an unfertilized egg. There is a large gray area in between, particularly in the time between fertilization and birth. Some believe a fertilized egg, which in the correct environment has the potential for independent life, should be granted full moral status equivalent to that of an independently living and breathing human being. Others disagree, believing an embryo should not be granted this status until it reaches later stages of development.

Many seek guidance on these complex issues from their religious traditions. For this reason, it is instructive to examine what the world's major religions have to say about human embryonic stem cell research.[12] Not surprisingly, given the controversy the topic inspires, there are significant differences in beliefs both between and within religions.

Christian beliefs regarding human embryonic stem cell research and therapeutic cloning vary among the religion's many branches. Catholic theologians almost universally reject embryo research, arguing that from the moment of conception a human embryo has a well-defined identity and deserves the full rights granted to any human being. Although this belief follows the Catholic Church's position on abortion, it does not necessarily reflect the church's historical views. From the time of Saint Augustine to the late nineteenth century, official church teaching maintained that an unformed fetus lacked a human soul. Given this history, embryo research and perhaps therapeutic cloning, had they been possible earlier, may not have been viewed as negatively by the Catholic Church as they are today. The Eastern Orthodox tradition has expressed a similar view, rejecting all human embryo research. In contrast, many Protestant traditions hold more permissive views. The United Church of Christ, for instance, has publicly supported research on human embryos up to the fourteenth day of development.[13]

Jewish religious traditions are generally supportive of embryo research. In Judaism, moral status is not typically ascribed to a developing embryo until forty days after fertilization. This window provides ample time for the derivation of human embryonic stem cell lines from human blastocyst-stage embryos and permits therapeutic cloning research. The Jewish tradition also emphasizes the importance of saving lives. Some theologians have argued that the noble end of human embryonic stem cell research justifies any negatives associated with the use of early embryos in research. The situation is similar in Islam, which most scholars and religious leaders agree permits the derivation of human embryonic stem cell lines from early human embryos. In a statement presented to the U.S. National Bioethics Advisory Commission, a Muslim religious scholar summarized the beliefs of various Islamic traditions as: "the fetus is accorded the status of a legal person only at the later stages of its development, when perceptible form and voluntary movement are demonstrated."[14] This assessment is borne out by the successful derivation of human embryonic stem cell lines in Iran in 2003.

Buddhist scholars report diametrically opposed views of embryonic stem cell research. Some view the ends of the research as the crucial factor: if the intention is the improvement of human health, research is permissible but otherwise embryo research should be forbidden. Other Buddhist scholars have argued that all embryo research should be prohibited, since the First Precept of Buddhism prohibits causing death or injury to living creatures.

This debate over the assignment of moral status to a developing embryo defines most people's views on the ethical acceptability of human embryonic stem cell research and therapeutic cloning. Those who believe that moral status should be granted at fertilization for the most part oppose the research. Those choosing later events, such as implantation, the development of a nervous system, the first perceptible movement, or birth to assign moral status to a developing human embryo are typically more supportive.

The egg debate

A second ethical debate surrounding therapeutic cloning research relates to its use of human eggs. Because therapeutic cloning requires the transfer of genetic material from an adult cell into an enucleated egg, each attempt requires a human egg. This requirement acts as a constraint on research because mature human eggs are difficult to acquire. As we saw earlier, they are retrieved surgically from women who have undergone hormone injections. The process can be uncomfortable and has potential side effects, including ovarian hyperstimulation syndrome, a condition marked by fluid buildup in the chest and abdomen and enlargement of the ovaries. Although most cases are mild, ovarian hyperstimulation syndrome can lead to life-threatening complications.[15]

Bioethicists worry that therapeutic cloning research, by increasing the demand for human eggs, will increasingly lead scientists to take shortcuts when recruiting egg donors. As we saw in the South Korean research, coercion is a concern. In general, donations by junior members of research teams, such as graduate students, post doctoral researchers, or research staff are considered inappropriate. Although these scientists may genuinely want to donate eggs to advance research, the risk of coercion is high and the appearance of it inevitable. Theoretically, these concerns can be reduced by limits on who is allowed to donate but the great pressures associated with modern science, in which many labs are competing to publish their findings as quickly as possible, mean they cannot completely be eliminated. A related concern is that research donors may not fully understand the risks associated with their donation. This concern can be addressed through informed consent procedures in which participants are told about the research, their role in it, and the risks involved. However, the effectiveness of informed consent varies and compliance is difficult to enforce.

Compensation is yet another potential concern. Egg donors for *in vitro* fertilization are routinely compensated in the United States and some ethicists worry that high levels of compensation can potentially lead women to donate against their best interest. Most countries suggest that egg donors for embryonic stem cell research should not be compensated (beyond their medical expenses and costs clearly associated with the donation) but the chance exists that if donors are hard to find, compensation will be offered. This risk can be seen in the payment of egg donors in South Korea. It is not yet clear how difficult the recruitment of egg donors for therapeutic cloning research will be. As research advances, eggs may become a limiting factor and bring these ethical concerns to the fore.

Further reading

Readers seeking additional information on human reproductive cloning and the ethical arguments for and against it have numerous options. Lee Silver, in his book *Remaking Eden: How Genetic Engineering and Cloning will Transform the American Family* (Harper Perennial, 1998), published shortly after Dolly's birth, presents far-ranging and provocative views on the potential impact of cloning and other reproductive technologies.

Gregory Pence, a bioethicist, presents a variety of interesting arguments in favor of human reproductive cloning in his 2005 book, *Cloning after Dolly: Who's Still Afraid?* (Rowman & Littlefield Publishers, 2005).

Contrasting views are available in *Life, Liberty and the Defense of Dignity: The Challenge for Bioethics* (Encounter Books, 2002), by Leon Kass, chair of the President's Council on Bioethics in the United States from 2001 to 2005.

Detailed overviews of the arguments both for and against human cloning can be found in the reports of the two major U.S. bioethics commissions that have considered human cloning. The first report, *Cloning Human Beings*, was produced by the National Bioethics Advisory Commission in 1997 and focused solely on human reproductive cloning. The second report, produced in 2002 by the President's Council on Bioethics, a new committee formed to advise President Bush, was titled *Human Cloning and Human Dignity: An Ethical Inquiry*. It addressed issues raised by both reproductive and therapeutic cloning.

Background on the science of human cloning and the steps scientists must overcome before human reproductive cloning might be safe enough to consider can be found in *Scientific and Medical Aspects of Human Reproductive Cloning* (National Academies Press, 2002).

For those interested in placing the debate over embryo research and human embryonic stem cell research in a historical context Jane Maienschein's *Whose View of Life: Embryos, Cloning and Stem Cells* (Harvard University Press, 2003) is an invaluable resource.

7
Cloning science in uncertain times

The future of any new technology is uncertain and cloning, for a variety of reasons, faces a particularly precarious road. This situation results at least in part from the ethical controversy cloning engenders. Because politicians in different countries, and even different states within the same country, have reached different conclusions in the ethical debate, policies governing cloning research vary dramatically. These disparate policies may shape the field in unexpected and unusual ways.

Commercial uses of cloning technology also face an uncertain intellectual property environment. Numerous patents, owned by private companies and a bevy of academic scientists, cover various elements of the nuclear transfer technique and the derivation of human embryonic stem cells. How these patents are sorted out by various courts, and which are upheld or rejected in countries around the world, will influence the development of cloning technology, including the commercialization of cloned animals and the development of medical therapies based on therapeutic cloning.

An important and largely unanswered question is what impact these uncertainties are having on the development of cloning science. Some have claimed that the unusual regulatory patchwork governing therapeutic cloning and human embryonic stem cell research, in which neighboring countries may espouse diametrically opposed policies, is hindering their development. This may be because restrictions slow scientific progress in countries with large research communities. It may also be because policy differences lead to wasted energy and money, as policymakers work to lure scientists from one country to another and scientists find their research delayed as they close one laboratory and open another. These same dynamics apply to private companies that relocate in search of favorable policy environments or permissive intellectual property regimes. This chapter examines the various sources of uncertainty affecting the development of cloning science, with an eye on their potential impact on the field's development.

Cloning policy options

As soon as Dolly's birth was announced, governments around the world began to express interest in regulating cloning science. Within a week of the public announcement of Dolly's existence, President Clinton announced a ban on the use of federal funds for research on cloning human beings, calling on people to "resist the temptation to replicate ourselves."[1] Calls for a ban or regulation proceeded rapidly elsewhere. Yet, despite this initial governmental interest, the regulatory process has been anything but straightforward or uniform. Some countries relied on older rules, tailored not to cloning but to pre-existing technologies; others created stop-gap measures, outside the normal legislative process. In some, laws banning human cloning were passed with minimal delays but in others, successfully regulating cloning has proved almost an intractable challenge.

The blitz of policies introduced in the aftermath of Dolly's birth laid the groundwork for the diverse policies governing cloning science today. More recently, numerous countries have re-evaluated existing policies or crafted new ones in light of the hope inspired by therapeutic cloning and the fears inspired by the groups and individuals claiming to be cloning human beings.

Policymakers looking to regulate cloning (or almost any new technology) can select from several options, ranging from a complete ban on a technology's use to a *laissez-faire* approach in which the market dictates development. In between, governments can choose to institute a temporary ban (a moratorium) – a relatively restrictive approach, or allow the technology to be used within a regulatory framework – a relatively permissive approach. All these have been considered, by one jurisdiction or another, as policy options that could be applied to human cloning both for reproductive and therapeutic uses.

Prohibitions are a straightforward option. In theory, governments can pass laws outlawing the use of a given technology and set penalties designed to ensure compliance. In practice, defining the technology and enforcing the ban can prove challenging: for example, policymakers striving to ban human reproductive cloning must be careful not to limit unrelated technologies. The implications of proposed cloning laws have not always been fully considered. In the immediate aftermath of Dolly's birth, some laws that were considered (but thankfully never adopted) by the U.S. Congress would have banned not just human cloning but a variety of standard techniques used in biomedical research laboratories for nearly three decades.

Enforcing restrictions on human cloning poses numerous challenges primarily because cloning requires modest facilities and can be accomplished by a small number of people. Furthermore, assuming parents wanted to keep their use of cloning technology secret, detecting reproductive cloning after the fact would be challenging, if not impossible. For these reasons, prohibitions on human reproductive cloning would probably keep cloning out of the public sphere but may do little to prevent the pursuit of cloning by maverick scientists and parents desperate enough to break the rules. This is not to suggest that banning reproductive cloning is futile. Even if limited use is inevitable, a ban may reduce its likelihood and frequency.

These enforcement issues may be less of a concern for a ban on therapeutic cloning, since successfully developing human therapies based on therapeutic cloning will take many years and the openness of the biomedical research process provides a relatively straightforward monitoring mechanism. To acquire funding, scientists report their findings in publicly available peer-reviewed literature and enforcement officials could simply review this literature to identify relevant advances. In theory, well-funded scientists could work on therapeutic cloning in secret but given the scientific challenges, such a scenario is exceedingly unlikely. As we will see, bans on therapeutic cloning, though potentially enforceable, are more likely to result in scientists interested in the technology moving to more permissive research environments.

Numerous countries have enacted bans of various kinds on human cloning. Most, but not all, formally ban human reproductive cloning. Some, including Canada, France, Germany, Switzerland, and Taiwan, also ban therapeutic cloning, although in some cases these policies are under review and may change in the relatively near future. In the United States, federal rules ban the use of federal funding for therapeutic cloning research, while privately funded therapeutic cloning research is permitted in some states and prohibited in others.

A moratorium on cloning technology, either on reproductive or therapeutic cloning or both, poses similar definition and enforcement challenges. Other important questions include how long temporary restrictions should last and how they should be reviewed. Typically, policies enacting moratoriums call for a review of the science before restrictions are lifted. These temporary bans are often structured with what policymakers call a "sunset clause," meaning that without further legislation, the temporary ban automatically ends when its initial period is complete. Both Japan and the Netherlands have instituted moratoriums on therapeutic cloning research in recent years and several other countries have considered this policy option.

Allowing cloning to proceed under regulation encompasses a broad range of more permissive policy options. Regulation could be simple and allow human cloning for either reproductive or therapeutic purposes as long as certain safety requirements were met. Alternatively regulation could be extremely detailed and permit cloned embryos to be derived for research purposes only in a small set of precisely specified situations. Regulations could also, theoretically, be applied to human reproductive cloning. Assuming reproductive cloning was shown to be safe, it is possible to imagine a regulatory system in which use of the technology was permitted for heterosexual couples with irreversible infertility but not for other uses. At present, no countries explicitly permit human reproductive cloning but several, including the United Kingdom, Belgium, Israel, Singapore, South Korea, and China, permit therapeutic cloning research under varying degrees of regulation.

The *laissez-faire* approach is the most permissive policy option. It relies on human judgment and the functioning of free markets to serve society's best interests. While few countries are willing openly to embrace a free market approach to human cloning, limited regulation in some countries can be argued to represent an implicit acceptance of this strategy. The United States, where no federal legislation directly addresses human reproductive or therapeutic cloning, and where privately funded research faces few restrictions, is one example.

The therapeutic-reproductive cloning divide

Given the consensus against human reproductive cloning, it might seem that banning its use would be a relatively straightforward endeavor for legislators. In some countries, it has been. In others – most notably the United States – and in international debates, reaching agreements to ban reproductive cloning has proved challenging. This challenge has resulted not from disagreement over how to regulate reproductive cloning but from disagreements about therapeutic cloning.

If policymakers were willing to consider reproductive and therapeutic cloning separately, legislation to ban reproductive cloning would be passed in almost every country and multinational agreements could, in all likelihood, be reached with minimal delay. In theory, with this element of cloning technology addressed, policymakers could turn their attention to therapeutic cloning and decide if it should be banned, restricted, or encouraged. Yet this division of the cloning debate has been rejected by conservative lawmakers in the United States and by a coalition of countries, led by the United States and Costa Rica, in debates at the United Nations.

Separation of the cloning debate has been rejected primarily for two reasons. The first is political. While the consensus against reproductive cloning is strong, no such consensus exists against therapeutic cloning. Politicians opposed to all human cloning realize this and know that passing legislation to ban therapeutic cloning on its own would prove difficult. They hope, by grouping the two potential uses of human cloning together, to use the strong sentiment against reproductive cloning to help ban therapeutic cloning. Second, some policymakers reject separating the cloning debate because they fear that permitting therapeutic cloning puts society on a path that irreversibly leads to reproductive cloning. They argue that if scientists refine the techniques required to create healthy cloned embryos, these will greatly facilitate reproductive cloning. Furthermore, they suggest that if cloned human embryos exist in freezers at biomedical research laboratories, it will be only a matter of time until these embryos are transferred to the wombs of surrogate mothers, either deliberately or by mistake, potentially leading to the birth of a cloned human being. To avoid such risks, these policymakers maintain that all uses of cloning should be addressed by a single public policy.

The U.S. policy quagmire

This disagreement over how cloning policy should be addressed has reared its head most dramatically in the United States, where the policy environment can, at best, be described as haphazard. Attempts to pass federal cloning legislation ended in a stalemate: neither politicians striving to ban both reproductive and therapeutic cloning nor those wishing to restrict only reproductive cloning were able to muster enough votes. Cloning policy in the United States is further complicated by its link to the abortion debate. More than three decades after the United States Supreme Court granted women the right to terminate a pregnancy, abortion remains among the most divisive of policy issues and any policy debates that touch, even tangentially, on it generate intense partisanship and often prove intractable.

The result of this federal deadlock is an uncertain environment for cloning scientists. National rules restrict the use of federal money in human cloning research but do not limit privately funded research. The U.S. Food and Drug Administration (FDA), the federal agency charged with ensuring the safety of the nation's food supply and approving medicines, has, based on pre-existing statutes, claimed authority to oversee human reproductive cloning but some legal scholars believe this claim rests on a shaky legal footing. A patchwork of state cloning policies, addressing both human reproductive and therapeutic cloning, which has emerged in recent years, further complicates the situation.

The FDA's announcement that it intended to block any attempts at human reproductive cloning marks the closest the United States has come to a law on this controversial technology. However, the FDA's authority relies on rules created for other purposes and may extend only to a subset of potential human cloning experiments.[2] In particular, the FDA's mandate is safety, not ethics, and it is not clear the agency would have any grounds to limit human cloning, if it were shown to be safe. The FDA bases its authority on its role in regulating "biological products" – viruses, vaccines, blood, and the like – that are used to treat medical conditions. The agency argues that cloned human embryos are "biological products" created to treat infertility and thus fall under their jurisdiction. This argument suggests that the FDA could regulate cloning to treat infertility but not reproductive cloning undertaken for other non-medical purposes. The FDA has also claimed that it can regulate cloned human beings as "drugs." According to the FDA, drugs are "articles (other than food) designed to affect the structure or function of the body" and cloned human embryos qualify as articles that affect a woman's body, by making her pregnant. These claims have not been tested in court and whether they could withstand a legal challenge remains unclear.[3]

Therapeutic cloning research using private funding is not restricted at the national level but several states explicitly ban it. Thus, research that is encouraged in one state may be illegal in another. The use of federal funding is prohibited, forcing scientists interested in therapeutic cloning to look elsewhere for funding. This restriction dates back to a 1995 rule blocking the use of federal funds for any research in which human embryos are harmed or destroyed. This restriction, which as we saw in Chapter 5 forced James Thomson to turn to outside funding for his research isolating human embryonic stem cells, is significant, as the government is a major funding source for basic biomedical research conducted at academic institutions.

State cloning policies remain in flux. Several states, including California, New Jersey, Connecticut, and Massachusetts, explicitly permit therapeutic cloning research and some provide state funding. California – where voters approved three billion dollars in funding for stem cell research in November 2004 – seems poised to lead the way, although litigation has slowed the state's efforts. Therapeutic cloning is prohibited in at least six states and several others are considering bans. The majority of states have taken no action regarding this use of cloning technology, so in most of these states, therapeutic cloning is presumably legal.

Scientists working with human embryonic stem cells also face similar federal funding restrictions and a patchwork of state policies. Federal funding is limited to research on human embryonic stem cell lines derived from spare embryos before 9 August 2001. Just like scientists hoping to study therapeutic cloning, scientists wishing to study newer human embryonic stem cell lines must rely on private funding. Because of this, national U.S. policy toward human embryonic stem cell and therapeutic cloning research is generally considered restrictive.

The uniform U.K. system

The U.K.'s regulatory strategy toward human cloning contrasts with the U.S. system and warrants our attention. The United Kingdom, home to the world's first test-tube baby, has extensively regulated assisted reproductive technologies since 1990 and seemed well-positioned to oversee human cloning after Dolly's birth was announced in 1997. Research on human embryos is permitted for a small number of medical reasons; licensing and oversight is undertaken by the Human Fertilisation and Embryology Authority (HFEA), a regulatory body established by the British Parliament. Following the announcement of Dolly's birth, a review of existing policy concluded that HFEA rules effectively prohibited human reproductive cloning in the United Kingdom, but permitted therapeutic cloning.

This interpretation was challenged in court by the ProLife Alliance, a U.K.-based group opposed to abortion, embryo research, and human cloning, which argued that the HFEA rules applied only to embryos created through fertilization. This argument was upheld by the British High Court in November 2001, throwing the U.K.'s human cloning regulatory scheme into question. The ProLife Alliance presumably hoped this ruling would force the adoption of stricter rules banning human cloning for any purpose, but parliament dashed these hopes by taking quick action to ban reproductive cloning while re-establishing the legality of therapeutic cloning. This action left the United Kingdom with policies toward both human embryonic stem cell research and therapeutic cloning that are generally considered permissive. In contrast to the United States, these U.K. rules apply to all research, regardless of funding source.

Scientists wishing to create cloned embryos for research purposes must apply for a license from the HFEA and explain why the use of cloned embryos is critical and how the project fits an approved rationale for embryo research. The HFEA granted its first therapeutic cloning license to a team of scientists led by Alison Murdoch at the University of Newcastle upon Tyne in August 2004. Another license has since been granted to Ian Wilmut, who plans to use nuclear transfer to create embryonic stem cell lines from patients with motor neuron disease (amyotrophic lateral sclerosis). Wilmut and his collaborators hope this approach will help them understand the early stages of this neuro-degenerative disease, which has affected such notable personalities as the baseball player Lou Gehrig and the physicist Stephen Hawking.

The United Nations cloning debate

Given the different cloning policies adopted by individual nations, it is perhaps not surprising that debates on cloning policy at the United Nations (UN) have been controversial. Still, it is disconcerting that the UN cannot complete a binding multilateral treaty to ban human reproductive cloning, a technique that all member nations oppose. Much as in the United States, this policy has been blocked not because of support for reproductive cloning but because of disagreements over therapeutic cloning.

The UN cloning debate started in August 2001, when France and Germany – which ban all human cloning – proposed that the UN develop an international convention banning human reproductive cloning. This measure initially met with broad support and its passage was widely expected. This optimism, however, proved futile. The first significant hint of the challenges ahead occurred in early 2002, when the United States indicated that it thought both therapeutic and reproductive cloning should be addressed by the same resolution. This suggestion was opposed by countries that supported therapeutic cloning, as well as by France and Germany, who believed the UN should first focus on reproductive cloning, where there was a consensus, and turn to therapeutic cloning later.

In September 2003, a UN committee took up two proposals. One was a joint U.S.–Costa Rican proposal that banned human cloning for any purpose. The second was a Belgian proposal that banned only reproductive cloning. Debate was deadlocked. Finally a compromise of sorts, a two-year deferral, was reached and passed by a one-vote margin. Most countries supporting therapeutic cloning voted for the deferral, which was widely seen as a defeat for the United States and others hoping to ban human cloning for any purpose. Despite this, supporters of the U.S.–Costa Rican proposal continued to lobby for a ban on all human cloning; this effort led to a reduction in the deferral's length from two years to one.

When debate resumed a year later, little had changed. As it became apparent that agreement on a legally binding resolution could not be reached, Italy stepped in with another compromise, suggesting a non-binding declaration calling on countries to ban reproductive cloning and respect human dignity in research. The exact wording of the declaration was the source of much debate. Much of the appeal of the Italian compromise, beyond its non-binding nature, was that its language was ambiguous enough to permit both sides to claim victory.

The final wording of the Italian compromise called on UN member states to "prohibit all forms of human cloning inasmuch as they are incompatible with human dignity and the protection of human life."[4] Nations opposed to therapeutic cloning were free to claim that this technique was incompatible with human dignity and thus rejected by the UN resolution; countries supporting the technology could claim that research on such early embryos was compatible with human dignity or, because of the potential health benefits, important for the protection of human life, and thus acceptable.

The United Nations Declaration on Human Cloning was passed in March 2005 by an 84–34 vote, with 37 abstentions. Most non-binding UN declarations are passed unanimously but this declaration was opposed by many countries that supported therapeutic cloning research, due to concerns that the final wording was too broad and could be construed as banning the creation of cloned embryos for research purposes.

A permissive trend

Talk of a UN therapeutic cloning ban worried countries that supported this research. A binding resolution could have hindered existing or potential research in those nations. The same situation exists when individual countries re-evaluate their policies. Individual scientists studying therapeutic cloning or human embryonic stem cells run the risk that changes to existing policies could render their existing research illegal and force it to be abruptly halted.

Recent trends suggest this is unlikely. At a national level, most recent policy changes have been from restrictive to permissive policies. Several countries that initially chose to ban therapeutic cloning are reconsidering these policies and choosing to permit the use of nuclear transfer for research purposes. This is the case in

Japan where, following a moratorium on therapeutic cloning research, the government has announced its intention to authorize and support research in the field. It is also the case in Spain, where laws have come full circle in fewer than ten years. When human embryonic stem cells were first isolated in 1998, Spain banned all embryo research. Now the government has announced its support not just for research on spare embryos but also for those created using nuclear transfer technology. Even in the United States, where President Bush remains opposed to human embryonic stem cell and therapeutic cloning research, public opinion increasingly favors more permissive policies. Both branches of the U.S. congress passed a bill eliminating some of the restrictions on federal funding for human embryonic stem cell research, forcing the president to use his first veto in nearly six years in office.

If these trends continue, the environment for scientists studying therapeutic cloning may improve. However, policy change takes time and future changes may not necessarily follow this general trend. Cloning scientists seem likely to face a varied and uncertain policy environment for the foreseeable future.

The patent patchwork

Just as the regulations governing cloning research vary from country to country, the intellectual property environment also varies. This is especially true regarding the patenting of living organisms and patents related in some way to human life. Although significant progress toward harmonizing the patent systems used in various countries has been made in recent years, disagreements have persisted over patents on products, such as human genes, and processes, such as techniques to create embryonic stem cell lines, which touch on what it means to be human.

The patent system is designed to reward inventors. In exchange for publishing the details of their invention, inventors are granted a time-limited monopoly under which to commercialize their discovery. To qualify for patent protection, an invention must meet several criteria. In the United States, inventions must be novel, non-obvious, and useful, and the range of patentable material is broad. Although "products of nature" cannot be patented, if such products are removed from the natural environment, or processed in some way, they become "compositions of matter" and eligible for patenting. The U.S. Supreme Court, in *Diamond v Chakrabarty*, a landmark case permitting the patenting of a genetically modified bacterium, judged that patentability extended to "anything under the sun that is made by man."

This broad definition includes crucial cloning-related discoveries. Patents have been granted on elements of the somatic cell nuclear transfer technique to Ian Wilmut and others who worked at the Roslin Institute as well as other cloning scientists. Sorting out the claims made in these various patents is crucial for companies hoping to commercialize cloning technology. Disputes between some of these companies have already occurred. Geron and Advanced Cell Technology, two U.S. companies interested in cloning technology, have waged legal battles over their various and sometimes contradictory patents. After several years of litigation, the two companies settled their disputes in September 2006.

Intellectual property controversies also affect human embryonic stem cell research. Broad patents covering the derivation and use of human embryonic stem cell lines from primates, including humans, were granted to James Thomson at the University of Wisconsin. These patents, which are held by the Wisconsin Alumni Research Foundation, a non-profit-making affiliate of the University of Wisconsin, have been accused of slowing progress in the field. Geron, in exchange for funding Thomson's initial research, has been granted exclusive commercial rights to embryonic stem cell based therapies involving three promising cell types. Geron also owns the rights to the Roslin nuclear transfer patents, a combination which puts the company in a strong position to pursue therapeutic cloning research.

Embryonic stem cell patents face an uncertain future. They have recently been challenged in the United States and may be narrowed or declared invalid. Furthermore, although patents were granted for this technology in the United States, other nations have refused to grant them, claiming human embryonic stem cells are not patentable matter.

A policy-induced slowdown?

Assessing the impact of the diverse and unusual regulatory environment on the development of cloning technology is an important question that remains largely unaddressed. It is particularly relevant for therapeutic cloning which, due to the varied policy environment, can be pursued effectively in some countries but not others.

Some advocates are worried that restrictions on the creation of cloned human embryos for research purposes are hindering the development of novel therapies. Biomedical research progresses through an additive process, in which research groups pursue similar lines of research and build on each other's work. Cooperation and collaboration between research groups are important, as is, in many cases, competition. If large funding agencies, such as the U.S. National Institutes of Health, choose not to fund such research, and fewer laboratories pursue research, the rate of progress is almost certain to slow.

In therapeutic cloning research, this effect may be mitigated by unusually large investments made by countries that see it both as providing a unique opportunity to contribute to cutting-edge science and also perhaps to allow them to take a lead while other countries debate its ethical acceptability. The extent to which focused research efforts in countries including the United

Kingdom, China, and Singapore can compensate for reduced research in the United States and in other countries that oppose therapeutic cloning remains an open question.

Published studies suggest that a disproportionate share of research related to human embryonic stem cells is taking place outside the United States.[5] Compared to similar but less contentious biomedical technologies, the U.S. share of human embryonic stem cell publications is atypically low. Research addressing the rate of development of embryonic stem cell research worldwide has not yet been published.

Venue shopping

An additional concern is that this regulatory patchwork may be causing scientists to move from country to country in search of permissive regulatory environments, or "venue shopping." Policymakers in the United States are worried that their top scientists are leaving the country (or moving from restrictive states to permissive ones) to pursue their research. Policymakers in Europe are not immune from these fears either. Some restrictive European nations have voiced concerns about policy-induced brain drains, with scientists moving either to permissive European nations or to Asia.

Anecdotal evidence for these concerns exists. When Roger Pedersen, a top embryonic stem cell scientist, left the United States for the United Kingdom in 2001, he indicated that uncertainty over research policy in the United States played a crucial role in his decision. Both Alan Colman, one of the leaders of the team that cloned Dolly, and Laurence Stanton, a scientist at Geron, the company that funded James Thomson's initial work with human embryonic stem cells, have moved to Singapore. They moved because of the availability of research funding for their work in Singapore, rather than because of policy differences, but this may be indicative of a larger and less obvious migration of scientists to countries that support human embryonic stem cell and therapeutic cloning research. In recent years, the Chinese government has worked to convince some of the many Chinese citizens who have studied science abroad over the last two decades to return home; anecdotal reports suggest these efforts have been particularly successful in the fields of therapeutic cloning and embryonic stem cell research.

Because many scientists move without attracting media attention, it is hard to know the extent of such migration, but a survey of stem cell scientists in the United States suggests these scientists are disproportionately considering leaving the country to pursue their research.[6] Compared to similar biomedical scientists working in less-contentious fields, stem cell scientists were approximately five times more likely to have received a job offer for a position outside the United States within the previous year.

Such movement may not affect the rate of development of the whole field but it raises concerns for countries interested in the potential economic benefits of possible therapies. Just before the Hwang scandal broke, South Korea claimed that he held fourteen domestic and international patents covering the therapeutic cloning process, with seventy-one others under consideration at patent agencies around the world.[7] Although most, if not all, of these patents will probably be ruled invalid in the aftermath of the scandal, their existence, and the South Korean government's focus on developing intellectual property, highlights the attention some countries are paying to the potential economic benefits of cloning research.

It is also possible that companies could choose to "venue shop." Because the broad patents on human embryonic stem cell research apply in the United States but not in many other countries, the cost of doing this research, particularly for commercial firms, can be greatly reduced by leaving the United States. It is not clear if any company has or will leave the United States in search of a more favorable intellectual property climate but some multinational companies are reportedly considering shifting their human embryonic stem cell research to international subsidiaries.

Science responds to ethics

Beyond affecting the rate of research and the career plans of scientists, the prevalence of ethical concerns surrounding human embryonic stem cell research has led to an almost unprecedented situation in which some scientists are specifically undertaking projects to counter ethical objections. Two articles prominently published in *Nature* in 2005 illustrated different approaches to addressing concerns over the use of human embryos in research.

The first technique is "altered nuclear transfer." This is championed by William Hurlbut, a medical doctor and consulting professor at Stanford University, and is straightforward. Hurlbut proposes creating embryos that are modified such that they can never fully develop, perhaps by blocking the formation of a placenta. These "altered embryos" can be created by genetically modifying the donor cells used for nuclear transfer. Thus the genetic changes affect only cells in a petri dish, which everyone agrees have no significant moral status. After transplant, the embryo has no chance to develop fully and, in Hurlbut's view, does not deserve the moral status of a normal, unaltered, embryo. Hurlbut has presented this proposal in many venues and received relatively positive receptions from groups typically opposed to human embryonic stem cell research.

The proposal has generated harsh criticism from scientists and other bioethicists, however. Many, including some who grant more status to early human embryos, find the idea of creating what have been termed "handicapped" or "disabled" embryos even more repugnant than the standard research process for isolating embryonic stem cells. Regardless of this, scientists published a proof of principle paper in October 2005, verifying that altered nuclear transfer works in mice and could theoretically work in humans.[8] The team, which included the cloning expert Rudolf Jaenisch, genetically modified mouse cells such that a key gene required for placental development was disabled. They then used these cells as donors in a nuclear transfer experiment. The resulting cloned embryos did not implant successfully but embryonic stem cells were successfully isolated.

The second paper reported a technique for deriving human embryonic stem cells from mice without eliminating the embryo's potential for development. Rather than separating the inner cell mass from the trophectoderm by immunosurgery, the scientists, primarily from Advanced Cell Technology, removed single cells from eight-cell mouse embryos. These cells were cultured with previously isolated mouse embryonic stem cells genetically engineered to glow, which thus could readily be distinguished from the new cells. In some cases, the single isolated cells gave rise to new embryonic stem cell lines. The remaining seven-cell embryos were transplanted into female mice and developed normally to term. More recently, this approach was demonstrated to work with human cells.[9] The same team separated eight to ten cell human embryos and showed that these individual cells could give rise to human embryonic stem cell lines. This suggests that if a single cell was removed from an eight-cell human embryo, as is regularly done in a variant of *in vitro* fertilization, that cell could be used to create a human embryonic stem cell line without affecting the embryo's developmental potential.

Although this strategy offers no help to scientists contemplating therapeutic cloning (as nobody wants to transfer cloned embryos), it may reduce the controversy surrounding the derivation of human embryonic stem cell lines from spare embryos at fertility clinics. However, this strategy is open to criticism. In particular, the single cell removed from the developing embryo at the eight-cell stage may be developmentally equivalent to a fertilized egg and, if given the correct conditions, could potentially give rise to a healthy organism. If this single cell is equivalent to an embryo, the method offers little to quiet critics opposed to the use of embryos in research.

These publications are noteworthy, not so much for their science as for their role in ongoing ethical debates. In this regard, their impact has been mixed. They have not by any means ended these debates, but they have played rather prominent roles, primarily by providing a third option for U.S. lawmakers uncomfortable with current federal funding restrictions but unwilling to support research on newer human embryonic stem cell lines. Recent political debates in the United States have included discussions of increased funding for these alternative methods of deriving human embryonic stem cell lines, and some lawmakers support this rather than a policy overturning some of the existing restrictions.

In the long term, many scientists are hopeful that patient-matched pluripotent stem cells can be developed without harming embryos and without these sorts of awkward workarounds. They hope that, by understanding the genes that define a pluripotent state, they will be able to convert adult cells directly to cells that exhibit embryonic stem cell-like properties. This is a long-term project that will rely on an understanding developed through studies of human embryonic stem cells but it may eventually reduce tensions over research in this field and lessen the impact of the unusual policy patchwork currently governing it.

Further reading

Additional information on how public policy can, and perhaps should, address cloning science can be found in several books. Among the best are *Crafting a Cloning Policy: From Dolly to Stem Cells* (Georgetown University Press, 2002). This book, by Andrea Bonnicksen, a political science professor at Northern Illinois

University, focuses on policy development in the United States, providing both a history of policy toward embryo research and chronicling congressional policy debates in the late 1990s. It also examines policy approaches taken in the United Kingdom, Canada, and Australia.

Another option is *Human Cloning: Science, Ethics and Public Policy* (University of Illinois Press, 2000). This book, edited by Barbara MacKinnon, contains a series of essays by leading scholars examining how policy could address cloning technology under a variety of assumptions. Policy issues raised by cloning are addressed as well in *The cloning sourcebook* (Oxford University Press, 2001), a compilation of 27 essays edited by Arlene Klotzko. This useful book also includes sections on the science and ethics of cloning.

For readers interested in the intersection between the law and human cloning, Kerry Lynn Macintosh's *Illegal beings: human clones and the law* (Cambridge University Press, 2005) provides a thorough assessment of the legality (or illegality) of laws restricting human cloning contemplated or on the books of various U.S. jurisdictions.

Because cloning policy remains in flux, particularly in regard to therapeutic cloning or somatic cell nuclear transfer research, online resources provide a better source for keeping up to date with policy developments. Stem cell policy, including policy toward therapeutic cloning, is tracked by William Hoffman at the University of Minnesota. His World Stem Cell Map can be viewed at http://mbbnet.umn.edu/scmap.html.

Online references can provide additional insight into potential regulatory frameworks for human cloning research. Useful reports along these lines include *Regulating Human Cloning* produced by the American Association for the Advancement of Science and *Cloning: A Policy Analysis*, produced by the Genetics and Public Policy Center. These reports are available at the two groups' websites.

8

The future of cloning

Predicting the future of cloning technology is probably an exercise in futility. Our glance at the field's history shows how irregular and unpredictable major breakthroughs have been. The future of cloning science is further clouded by the controversy the technology generates and seems certain to be shaped by the large number of public policies that both support and restrict research. These policies may move research from one location to another, alter its pace, or even block it entirely.

The future of cloning may also be shaped in unpredictable ways by free markets responding to, or even creating, consumer demand. At the moment, there is little demand for human reproductive cloning and few scientists or private companies either able or willing to meet what little demand exists but this could change. Demand for fertility treatment was low when the first test-tube baby was born in 1978 but after this success, a market quickly emerged. Today, private clinics are a key force shaping the field, particularly in the United States, where the fertility industry remains largely unregulated.

In the short term, cloning research is driven by a few relatively clear aims. Goals for animal cloning include genetically engineering cows, pigs, sheep, and other agriculturally important animals so that they grow more rapidly, pollute less, and produce higher quality meat or milk. They also include converting some animals into biological factories, capable of producing medicines and other products at a fraction of the cost of traditional manufacturing approaches. For human cloning, the focus is on using cloning technology to understand a host of diseases better and in the longer term, on developing cellular replacement therapies based on patient-matched embryonic stem cells.

These goals meet with varying levels of acceptance around the world but each is accepted by a number of scientifically advanced nations and thus progress seems likely, if not inevitable. Cloning technology, as it advances, will open the door to a host of other longer-term possibilities, such as the genetic engineering and enhancement of human beings, but these possibilities typically meet with only limited acceptance and their future seems uncertain. Many of these possibilities are distant enough that multinational regulation, along the lines of the failed UN effort, has the potential to block their development or use. This chapter will elaborate on some of the nearer-term goals of cloning research and speculate on its potential longer-term impact.

Cloning and the food supply

Although products from cloned animals have not yet entered the food supply, the technology appears on the track to acceptance, at least in the United States, where genetic modification of plants is normal. A poll conducted in October 2005 suggested that, while many Americans are uncomfortable with animal cloning, a majority believe it has appropriate uses, including developing disease-resistant livestock, saving endangered species or gaining insight into human diseases.[1] Adoption of this technology seems less certain in Europe, where opposition to biotechnology in general, and genetic engineering more specifically, is more entrenched.

As we have seen, the use of cloning to duplicate valuable animals is rapidly advancing. If restrictions on bringing meat or milk from cloned animals to market are lifted, herds of cloned cows would in all likelihood begin to appear on many American farms, particularly if cloning efficiency improves. The commercial benefits of duplicating top milk-producers and prize-winning steers would simply be too much to resist. Moving from cloning existing animals to using the technology to produce genetically modified animals is the next logical, if somewhat uncertain, step. It is conceivable that discomfort with genetic modification could delay or block this, but it seems more likely that market forces will prevail and it will gain acceptance in the United States and, assuming no serious complications arise, around the world.

Take mad cow disease as an example. Only three cases have been reported in the United States but estimates suggest the disease cost the U.S. beef industry up to $4.7 billion in 2004, mainly because of restrictions on the export of beef. This cost is dwarfed by the expense of the twenty-year long epidemic in the United Kingdom. If, as some research suggests, cows could be reliably genetically modified to resist this disease, this may well be deemed acceptable. Furthermore, the economics of such genetic modification may make sense not just for individual farmers but for whole nations. It is conceivable that if such protection could be introduced efficiently, governments might choose to encourage, or even require, all cattle destined for human consumption to carry the genetic modification. Such a policy may well give a country economic security, by protecting its export market and meet with relatively little public opposition, as few people have problems with protecting animals from disease.

Other modifications may prove more controversial, especially genetic engineering that seems to cross species boundaries in unnatural ways, such as the recently reported advance to produce so-called "heart-healthy" pigs. Omega-3 fatty acids, found primarily in oily fish such as salmon, tuna, and trout, have been linked with a reduced incidence of cardiovascular disease, yet many people don't eat enough fish to consume sufficient quantities of these beneficial fatty acids. Cloning could provide one solution. Scientists at the University of Pittsburgh and the University of Missouri-Columbia recently created cloned pigs that had abnormally high omega-3 fatty acid levels.[2] Because pork is regularly consumed by more of the world's population than is fish, these cloned pigs (or their offspring) might increase omega-3 fatty acid consumption worldwide. Will this modification prove controversial? It is hard to say, but the methodological details might make some pork-eaters cringe: the scientists inserted a gene from a common flatworm into the pig's genome. Although there is no reason to expect meat from these pigs to taste wormy, it is not hard to imagine consumer backlash against this pig/worm combination.

Animal factories

Cloning technology is also advancing the vision of creating genetically modified animals that produce valuable biological molecules. The economics of this model vary from drug to drug. Some proteins that are expensive to produce and purify in the laboratory seem quite likely to be produced more efficiently in genetically modified animals or plants in the future. Other, simpler, compounds may always be produced in more traditional manufacturing facilities.

The key to the economic efficiency of these animal factories is that almost all of the cost is up-front. To take a hypothetical example, successfully modifying a cow to produce insulin in its milk poses significant challenges and requires large amounts of capital but once a small herd of these cows exists they produce insulin at low cost. Indeed, they can be left to graze, essentially living off the land. Furthermore, they self-replicate. When cows from the genetically modified herd mate, their offspring should also produce insulin. This means that once the first few insulin-producing cows had successfully been created through cloning, many more could easily be produced by traditional breeding. The only remaining significant cost would be isolating the insulin from the cow's milk.

This strategy easily allows production to be decentralized so these "pharming" strategies have been suggested for handling some of the serious health problems facing Africa and other regions with relatively underdeveloped health systems. The idea is simple. Pharmaceutical-producing animals or plants could be given to communities in which access to important vaccines or medications is limited. The communities could then take responsibility for tending the plants or caring for the animals and harvest and use the biological products produced to improve the health of their population. Whether such a vision will ever come to pass is an open question. Research is just beginning and implementation remains distant. There is some hope that the strategy may prove successful, particularly as a low-cost means of serving markets that drug companies might otherwise ignore: in 2004, scientists in a South African laboratory announced that they were refining techniques to use genetically modified tobacco plants to produce drugs to treat HIV and tuberculosis.

Despite its potential benefits, this research remains controversial. This is particularly true in the United Kingdom and Europe, where genetic modification of plants and animals for any reason has met with only limited acceptance. Anti-GM groups claim the technology is unsafe for a variety of reasons, including the potential for gene transfer between modified and wild variants, perhaps leading enhanced variants to crowd out native species and reduce biodiversity.

The primary concern expressed about the use of genetically modified plants or animals to produce therapeutic compounds is that these enhanced organisms might interbreed with unmodified organisms and contaminate the food supply. In theory, drug-producing animals could be isolated from unmodified herds, preventing interbreeding, but opponents of the technology worry that isolation may not prove feasible and that contamination is inevitable. Blocking interbreeding between modified and unmodified plants poses greater challenges, since airborne pollen can travel long distances, but a number of strategies to control or limit interbreeding exist.

Opponents of genetic modification, whether or not it relies on cloning, have in some cases resorted to violence to make their point.[3] Fears that field trials in the United Kingdom would be disrupted by vandals were one reason that British research into producing a rabies vaccine in genetically modified corn or tobacco was relocated to South Africa. Although the economic benefits of this technology seem likely to drive continued development, the chance exists that research will grind to a halt in those countries where opposition to genetic enhancement is strong. In stark contrast to the situation in Europe, genetic modification has met with relatively little resistance in the United States. Because of these differences research, and particularly field testing of animals modified to produce therapeutic compounds, seems likely to be centered in the United States.

Cloning and medical research

While market forces seem almost certain to drive animal cloning research forward, whether cloning technology will have a significant impact on medical care is less clear. Many scientists are hopeful that stem cells derived from cloned embryos will be crucial to a host of novel therapies. Medical research is notoriously unpredictable, however, and it is possible that, despite these high hopes, the therapies may not emerge. Of course, it is also possible that the scientists' hopes will be fulfilled, if not exceeded.

Although no consensus has been reached regarding this research, as the UN cloning debate illustrated, there is a growing trend toward acceptance of research using embryonic stem cells. This is a necessary first step to acceptance of therapeutic cloning. Although politicians may try to distinguish them, encouraging research on human embryonic stem cells is essentially a tacit endorsement of the future creation, through therapeutic cloning, of patient-matched embryonic stem cell lines. This is because it would likely prove politically untenable to support research to the point where scientists show that human therapies are feasible but to refuse to support the one last step that would allow the therapies to be tested and used.

The trend toward acceptance of research along these lines, coupled with strong support in the United Kingdom, Sweden, Singapore, and China, among other countries, suggests that research will continue and advance. Research will no doubt be hindered by funding restrictions in the United States and other countries but while these restrictions might slow discoveries or shift research from one country to another, they seem unlikely to block it entirely.

These considerations suggest that, given time, therapeutic cloning science will get a chance to advance and, if feasible, lead to human therapies. The time such research will require is an important unknown. If the field advances slowly or, as is almost certain, hits unanticipated obstacles, opponents may claim it doesn't deserve support and that funding would be better directed to less controversial research areas. These sorts of claims have been common following the revelations of fraud in the South Korean therapeutic cloning reports.

The bigger question may be how the field will evolve if some therapies are developed. What if cell-based therapies prove useful for treating some diseases but whole organ transplant holds greater promise for others? Some have speculated that scientists or doctors could be tempted to allow cloned embryos to survive past the blastocyst stage, allowing development to proceed to the point where immature organs could be harvested and perhaps matured outside the body for eventual transplant.[4] This would mean significant advances in the development of artificial wombs but would not necessarily require the technology to develop a fetus to term outside the body. Hung-Ching Liu, an embryologist at Cornell, has indicated, in unpublished research, that she has grown mouse fetuses for seventeen days in artificial wombs.[5]

Seventeen days is only four days short of full-term for a mouse pregnancy. Although the mice were dead when she removed them, they had developed significantly and, importantly, had organs that may have been suitable for transplant. Extrapolating this research to humans suggests that similar technology, if it were considered ethically permissible, might allow human development to proceed to almost thirty-one weeks *in vitro*. This hint of success using mice should not suggest that true artificial human wombs are likely in the near future. Few scientists are actively working in this area and many challenges remain.

This scenario of cloned humans as organ factories raises numerous ethical questions and is, at the current time, opposed by the vast majority of scientists and ethicists who have considered it. However, it is one way in which this technology could develop in the distant future. Although such use of human fetuses seems unlikely today and a law banning so-called fetal farming was recently enacted in the United States, it is hard to say how society may decide to balance potential health benefits with harm to human embryos or fetuses in the future.

An alternative approach may rely on embryonic stem cells created through therapeutic cloning. Theoretically, these early cells could be coaxed into precursor cells of the organ of interest, which could further develop *in vitro* and be placed on a scaffold of some sort that promoted further development. Research is advancing on three-dimensional matrices that may eventually support organ development, but the challenges posed by recreating the organ development process in a dish are daunting.

Cloning, genetic engineering, and the future of humanity

Further in the future is the possibility that humans will one day take advantage of cloning technology to alter what it means to be human. The idea, though frightening to many, is relatively simple. Just as cloning is coupled with genetic enhancement in animals, it could be applied to human beings. In theory, changes could be made in human somatic cells, for example skin cells, in a petri dish and only the successfully modified cells used as the donor cells for somatic cell nuclear transfer. It would thus perhaps be possible to create genetically modified human beings. Importantly, humans produced through this procedure would have genetically modified germ cells and pass on their genetic changes to their offspring. For this reason, the technique is known as germ-line genetic engineering.

Although this procedure uses cloning technology, it does not necessarily mean that every genetically modified human would be genetically identical to his or her parent. Rather, it seems more likely that the technique would be applied to early embryos and thus each child would contain genetic material from two parents. Doctors would isolate cells from these embryos and genetically modify them for use as donors for nuclear transfer procedures.

The science required to accomplish such a feat is not trivial, but there is no reason to believe it is impossible. Indeed, scientists have every reason to believe that if research along these lines was pursued, it would prove technologically feasible. Each of the required steps has been accomplished in animals. Molecular biology has shown us, time and again, that if something can be done in mice, it can be done, given enough time, in humans. However, little research is proceeding along these lines, and it is not clear that society wants it to proceed.

The question of what society wants may be particularly relevant. Unlike some of the other uses of cloning technology, there is no current developed market clamoring for germ-line genetic modification. Such a market may appear, as some parents will spend almost limitless sums to improve their children's chances in life, but society could choose to regulate or restrict this technology.

Restrictions on human reproductive cloning are one strategy to hinder this particular approach to genetically modifying humans. Despite difficulties enacting policies and treaties to ban reproductive cloning, a global consensus against its use exists. If this holds and scientists choose to follow the wishes of their governments, it is conceivable that humans will not be cloned for reproductive purposes. Since this protocol relies on nuclear transfer, such a ban would hinder its development.

Many people feel that human reproductive cloning is inevitable. Despite the consensus against its use, a couple of doctors and an infertile couple who chose to ignore public opinion and pursue the technique could create the first cloned human. And once the first cloned child is born, it may, as with *in vitro* fertilization, legitimize the technology and pave the way for a rapid and dramatic change in public opinion. Continuing research toward therapeutic cloning also increases the likelihood of successful reproductive cloning. Once scientists have refined techniques for creating cloned human embryos and growing them to the blastocyst stage, it will be relatively straightforward for would-be reproductive cloners to follow the methods outlined in published therapeutic cloning reports and create potentially viable cloned embryos for reproductive purposes. These considerations suggest that restrictions on human reproductive cloning might prove insufficient to prevent human germ-line genetic modification.

Another option for opponents of this technology is to ignore cloning but argue directly against the introduction of genetic modifications into human germ cells or their precursors. Any donor cell used for nuclear transfer is by definition a germ cell precursor, since it gives rise to the entire organism. A general consensus against these heritable modifications exists and it might be possible to enact a worldwide ban against the procedure. A Council of Europe convention already blocks germ-line gene modification, and although it has been signed by only a minority of European nations, could be a template for wider multinational regulation.

Much as with cloning, enforcing such a ban could prove difficult. Genetic engineering requires only a few scientists and is almost impossible to detect. Thus, it is equally possible to imagine that attempts to block the use of this technology will falter. And if human germ-line genetic modification was safely used, it may well have a self-reinforcing effect. Imagine that you were a potential parent. You knew that other parents were using this technology to have children that, due to their genetic enhancement, were smarter, healthier, or improved in some way. Would you want to use this technology too, to give your child every possible advantage? Advances in cloning and related fields could force just this sort of dilemma on potential parents in the future.

To understand this dilemma, we need to examine why parents might be interested in this sort of technology. What benefits could genetic modification of humans beings bring? It is difficult to know for sure but the possibilities are nearly boundless. In theory, any gene from any organism could be added to the human genome. Most would have no effect, many would be harmful, but some might lead to humans with new and unique abilities. Genetic modification need not be limited to a single gene: a genetically modified person could have numerous genetic changes or even, as some have suggested, entirely novel chromosomes.

It seems likely that the first changes would be health-related. Who wouldn't want an extra copy of a gene variant that offered protection against Alzheimer's disease? What about a set of genes that blocked the development of cancerous tumors? Or a gene that coded for a protein which prevented the HIV virus from entering cells? While these are presently hypothetical examples, scientists know that genes play important roles in Alzheimer's and many types of cancer and have been studying how the HIV virus infects individual cells. This research may open the door to treating these and other diseases through genetic modifications at some point in the future.

Assuming genetic modification met with some acceptance, it seems likely that its focus would move from health to a broader range of enhancements. Genetic engineering might improve muscle tone, visual acuity, memory, or height, to name just a few examples. Many of these traits are both generally desirable and at least partly genetically determined. The question is whether desirability is a good enough reason to alter genes.

Some ethicists and scientists have suggested that society draw a line between medical therapies and enhancement, arguing that genetic modification may be acceptable for health reasons but not enhancement. This line, however, may not always be clear. Sometimes short stature is associated with a deficiency in human growth hormone and children with it can be treated with hormone injections. Yet other short children do not suffer from this deficiency. These children as well as children of normal height – or their parents – may feel that a few extra inches would be beneficial and want the treatment as well. Is the first case therapy and the latter enhancement? Or do they both represent enhancement? Real-world experience suggests it is hard to draw a line separating the two. The same would almost certainly be true for genetic engineering.

The long-term implications of such technology are impossible to predict. A world in which genetic engineering was common would be very different than the world we know today. Reproduction would shift from the bedroom to the clinic. Certain genetic traits might become more prevalent while others might decline. Perhaps these changes would be for the best. People might live longer and healthier lives, as common illnesses were conquered. Human abilities might reach new extremes as those engineered for athletic ability, greater intelligence, and so on broke previous boundaries.

Alternatively, acceptance of this technology could have negative consequences. It might split society even more rapidly into haves and have-nots. Genetic modification would no doubt be expensive at first and in the absence of measures to ensure its affordability, it might be limited to the wealthy. The combination of cloning technology and genetic engineering might accentuate already significant differences between social classes, perhaps leading to a permanent schism between different segments of society. Genetic enhancement might increase tension between parents and children. Conceivably, parents might have unreasonable expectations for their engineered or "designer" children. Conversely, children might hold their parents even more accountable than they do today for their abilities, successes, and failures.

This future is far from certain. Political leaders, scientists, and people around the world must decide if they accept the application of cloning and genetic engineering to humans and how best to regulate its use.

Cloning choices

This is an exciting time for cloning science. In many ways, the future looks bright. Research is clearly advancing in some areas and is poised to advance in others. But its future is only partially determined. Key advances have been made but many discoveries and challenges remain.

Governments are discussing and debating how this science should be pursued. Society faces difficult questions, as it balances the wide-ranging benefits of cloning science against its costs. These challenges are multiplied by the uncertainty inherent in the field. Scientists are unsure how likely therapeutic cloning is to cure diseases. Nor are they certain how long it will take, or if it is even possible, to eliminate the health concerns that affect cloned animals and are slowing the application of cloning technology.

Debates over the future of cloning technology should be grounded on a solid understanding of the underlying science, which allows these discussions to proceed logically with clear bounds on what is and is not feasible. My hope for this book is that it will help you understand this exciting science and participate meaningfully in the debates that will shape its future.

Further reading

Numerous books examine the potential future impact of cloning and other genetic engineering technologies. A noteworthy recent addition to this collection is *After Dolly: The Uses and Misuses of Human Cloning* (W.W. Norton and Company, 2006) by cloning pioneer Ian Wilmut and science writer Roger Highfield. In this book, Wilmut argues that scientists should one day be allowed to combine cloning and genetic modification to alter the human germ-line.

Another interesting perspective is provided in *Redesigning Humans: Our Inevitable Genetic Future* (Houghton Mifflin, 2002) by Gregory Stock, Director of the Program on Medicine, Science, and Society at the University of California – Los Angeles School of Medicine. As the title suggests, Stock argues that genetic enhancement is inevitable and that, in time, human society will come to embrace it much in the way it has embraced *in vitro* fertilization.

Lee Silver's *Remaking Eden: How Genetic Engineering and Cloning will Transform the American Family* (Harper Perennial, 1998) is noteworthy for its portrayal of a future society splintered, by cloning and genetic engineering, into two distinct classes, the GenRich and the Naturals. Silver provocatively suggests that over time, these two classes of humans could theoretically diverge and develop into separate species.

Arguments in opposition to human genetic engineering are provided by Francis Fukuyama, a member of the President's Council on Bioethics in the United States, in *Our Posthuman Future: Consequences of the Biotechnology Revolution* (Profile Books, 2002) and Bill McKibben in *Enough: Staying Human in an Engineered Age* (Times Books, 2003).

Notes

What cloning is and why it matters

1 Justin Gillis, "Clone-Generated Milk, Meat May Be Approved: Favorable FDA Ruling Seem as Imminent," *Washington Post*, 6 October 2005, p. A1

A cloning parts list: cells, genes, and embryos

1 James D. Watson and Francis H.C. Crick, "Molecular Structure of Nucleic Acids," *Nature* 171(4356), 1953, pp. 737–738.
2 Ying Chen *et al.*, "Embryonic Stem Cells Generated by Nuclear Transfer of Human Somatic Nuclei into Rabbit Oocytes," *Cell Research*, 13, August 2003, pp. 251–263.
3 Dagan Wells and Joy D. Delhanty, "Comprehensive Chromosomal Analysis of Human Preimplantation Embryos Using Whole Genome Amplification and Single Cell Comparative Genomic Hybridization," *Molecular Human Reproduction*, 6(11), 2000, pp. 1055–1062, and references therein.
4 Lewis Wolpert, *The Triumph of the Embryo* (New York: Oxford University Press, 1991), p. 12.

Dolly and her scientific predecessors

1 Gina Kolata, "Scientist Reports First Ever Cloning of Adult Mammal," *New York Times*, 23 February 1997, p. 1.
2 Hans Spemann, *Embryonic Development and Induction* (New Haven: Yale University Press, 1938), p. 211.
3 Robert G. McKinnell, *Cloning: A Biologist Reports* (Minneapolis: University of Minnesota Press, 1979), p. 21.

4 Marie A. Di Berardino, *Genomic Potential of Differentiated Cells* (New York: Columbia University Press, 1997), p. 35.

5 Robert Briggs and Thomas J. King, "Transplantation of Living Nuclei From Blastula Cells into Enucleated Frogs' Eggs," *Proceedings of the National Academy of Sciences*, 38(5), May 1952, pp. 455–463.

6 Gina Kolata, *Clone: The Road to Dolly and the Path Ahead* (New York: William Morrow and Company, 1998), pp. 65–66.

7 John B. Gurdon, "Adult Frogs Derived from the Nuclei of Single Somatic Cells," *Developmental Biology*, 4, April 1962, pp. 256–273.

8 Di Berardino, *Genomic Potential*, p. 76.

9 David Rorvik, *In His Image: The Cloning of a Man* (Philadelphia: J.B. Lippincott Company, 1978).

10 Karl Illmensee and Peter C. Hoppe, "Nuclear transplantation in Mus musculus: developmental potential of nuclei from preimplantation embryos," *Cell*, 23(1), pp. 9–18.

11 Kolata, *Clone*, pp. 130–134.

12 Kolata, *Clone*, p. 137.

13 Debora MacKenzie, "Illmensee inquiry finds chaos—but no fraud," *New Scientist*, 23 February 1984, pp. 3–4.

14 James McGrath and Davor Solter, "Inability of Mouse Blastomere Nuclei Transferred to Enucleated Zygotes to Support Development in Vitro," *Science*, 226(4680) 14 December 1984, pp. 1317–1319.

15 Steen M. Willadsen, "Nuclear Transplantation in Sheep Embryos," *Nature*, 320(6057), 6 March 1986, pp. 63–65.

16 Randall S. Prather *et al.*, "Nuclear Transplantation in the Bovine Embryo: Assessment of Donor Nuclei and Recipient Oocyte," *Biology of Reproduction*, 37(4), November 1987, pp. 859–866.

17 Ian Wilmut, Keith Campbell, and Colin Tudge, *The Second Creation: Dolly and the Age of Biological Control* (Farrar, Straus and Giroux, 2000), p. 152.

18 Keith H. Campbell *et al.*, "Nuclear-Cytoplasmic Interactions during the First Cell Cycle of Nuclear Transfer Reconstructed Bovine Embryos: Implications for Deoxyribonucleic Acid Replication and Development," *Biology of Reproduction*, 49(5), November 1993, pp. 933–942.

19 Keith H. Campbell *et al.*, "Sheep Cloned by Nuclear Transfer from a Cultured Cell Line," *Nature*, 380(6569), 7 March 1996, pp. 64–66.

20 Ian Wilmut *et al.*, "Viable Offspring Derived from Fetal and Adult Mammalian Cells," *Nature*, 385(6619), 27 February 1997, pp. 810–813.

Animal cloning in the twenty-first century

1 Teruhiko Wakayama *et al.*, "Full-Term Development of Mice from Enucleated Oocytes Injected with Cumulus Cell Nuclei," *Nature*, 394(6691), 23 July 1998, pp. 369–374.

2 Byeong Chun Lee *et al.*, "Dogs Cloned from Adult Somatic Cells," *Nature*, 436(7051), 4 August 2005, p. 641.

3 Calvin Simerly *et al.*, "Molecular Correlates of Primate Nuclear Transfer Failures," *Science* 300(5617), 11 April 2003, p. 297.

4 Wilmut, "Viable Offspring."

5 Wakayama, "Full-Term Development."

6 Jose B. Cibelli *et al.*, "The Health Profile of Cloned Animals," *Nature Biotechnology*, 20(1), January 2002, pp. 13–14.

7 David Humpherys *et al.*, "Abnormal Gene Expression in Cloned Mice Derived from Embryonic Stem Cell and Cumulus Cell Nuclei," *Proceedings of the National Academy of Sciences*, 99(20), 1 October 2002, pp. 12889–12894.

8 Jonathan E.M. Baillie, Craig Hilton-Taylor, and Simon N. Stuart (Editors), 2004 *IUCN Red List of Threatened Species: A Global Species Assessment* (IUCN: Gland, Switzerland and Cambridge, 2004).

9 Naomi Aoki, "Worcester Firm Touts Birth of Cloned Endangered Guar," *Boston Globe*, 12 January 2001, p. C1.

10 Pasqualino Loi *et al.*, "Genetic Rescue of an Endangered Mammal by Cross-Species Nuclear Transfer Using Post-Mortem Somatic Cells," *Nature Biotechnology*, 19(10), October 2001, pp. 962–964.

11 Humane Society of the United States. "Cat Cloning is Wrong-Headed States The Humane Society of the United States" 14 February 2002. (Press Release).

12 Humane Society of the United States. "HSUS Pet Overpopulation Estimates."

13 Matt B. Wheeler, "Production of Transgenic Livestock: Promise Fulfilled," *Journal of Animal Science* 81(Suppl. 3), 2003, pp. 32–37.

14 Angelika E. Schnieke *et al.*, "Human Factor IX Transgenic Sheep Produced by Transfer of Nuclei from Transfected Fetal Fibroblasts," *Science*, 278(5346), 19 December 1997, pp. 2130–2133.

Embryonic stem cells and the promise of therapeutic cloning

1 James A. Thomson *et al.*, "Embryonic Stem Cell Lines Derived from Human Blastocysts," *Science* 282(5391), 6 November 1998, pp. 1145–1147.

2 David I. Hoffman *et al.*, "Cryopreserved Embryos in the United States and their Availability for Research," *Fertility and Sterility* 79(5), May 2003, pp. 1063–1069.

3 Michael J. Shamblott *et al.*, "Derivation of Pluripotent Stem Cells from Cultured Human Primordial Germ Cells," *Proceedings of the National Academy of Sciences*, 95(23), 10 November 1998, pp. 13726–13731.

4 Yury Verlinsky *et al.*, "Human Embryonic Stem Cell Lines with Genetic Disorders," *Reproductive BioMedicine Online*, 10(1), January 2005, pp. 105–110.

5 William M. Rideout III *et al.*, "Correction of a Genetic Defect by Nuclear Transplantation and Combined Cell and Gene Therapy," *Cell*, 109(1), 5 April 2002, pp. 17–27.

6 Jong-Hoon Kim *et al.*, "Dopamine Neurons Derived from Embryonic Stem Cells Function in an Animal Model of Parkinson's Disease," *Nature*, 418(6893), 4 July 2002, pp. 50–56.

7 David Cyranoski, "Korea's Stem-Cell Stars Dogged by Suspicion of Ethical Breach," *Nature* 429(6987), 6 May 2004, p. 3.

8 David Cyranoski and Erika Check, "Stem-Cell Brothers Divide," *Nature* 438(7066), 17 November 2005, pp. 262–263.

9 Nicholas Wade "American Co-Author Wants His Name Off Stem Cell Paper," *New York Times*, 14 December 2005, p. A32.

10 Gretchen Vogel, "Picking Up the Pieces after Hwang," *Science* 312(5773), 28 April 2006, pp. 516–517.

The ethical debate over human cloning

1 Kenneth Chang, "Saying that Hoax is Possible, Journalist Leaves Cloning Test," *New York Times*, 7 January 2003, p. A12.

2 Joe Palca and Bob Edwards, "Human Cloning Efforts," Morning Edition, National Public Radio, 7 January 1998.

3 Panos M. Zavos and Karl Illmensee, "Possible Therapy of Male Infertility by Reproductive Cloning: One Cloned Human 4-Cell Embryo," *Archives of Andrology*, 52(4), July/August 2006, pp. 243–254.

4 Rick Weiss, "Human Cloning's 'Numbers Game'; Technology Puts Breakthrough within Reach of Sheer Persistence," *Washington Post*, 10 October 2000, p. A1.

5 President's Council on Bioethics, *Human Cloning and Human Dignity: An Ethical Inquiry* (Washington, DC: July 2002), p. xxvii.

6 U.S. Supreme Court. *Eisenstadt v Baird*, 405 U.S. 438 (1972) [Majority Opinion].

7 National Bioethics Advisory Commission. *Cloning Human Beings: Reports and Recommendations of the National Bioethics Advisory Commission* (Rockville, MD, June 1997) (Accompanying Material).

8 President's Council on Bioethics, *Human Cloning*, p. xxix.

9 President's Council on Bioethics, *Human Cloning*, p. xxviii.

10 National Consultative Ethics Committee for Health and Life Sciences. "Reply to the President of the French Republic on the subject of Reproductive Cloning," 22 April 1997.

11 Michel Revel, "Human Reproductive Cloning, Embryo Stem Cells and Germline Gene Intervention: An Israeli Perspective," December 2003.

12 National Bioethics Advisory Commission, *Cloning Human Beings*, pp. 39–61.

13 National Bioethics Advisory Commission, "Testimony of Ronald Cole-Turner," in *Ethical Issues in Human Embryonic Stem Cell Research, Volume III – Religious Perspectives* (Rockville, MD, June 2000), p. A-3.

14 National Bioethics Advisory Commission, "Testimony of Abdulaziz Sachedina," in *Ethical Issues in Human Embryonic Stem Cell Research, Volume III – Religious Perspectives* (Rockville, MD, June 2000), p. G-6.

15 The Practice Committee of the American Society for Reproductive Medicine, "Ovarian Hyperstimulation Syndrome," *Fertility and Sterility*, 82(Suppl. 1), September 2004, pp. 81–86.

Cloning science in uncertain times

1 Katharine Q. Seelye, "Clinton Bans Federal Money for Efforts to Clone Humans," *New York Times*, 5 March 1997, p. A13.

2 Rick Weiss, "Legal Barriers to Human Cloning May Not Hold Up," *Washington Post*, 23 May 2001, p. A1.

3 Richard A. Merrill and Bryan J. Rose, "FDA Regulation of Human Cloning: Usurpation or Statesmanship," *Harvard Journal of Law & Technology*, 15(1), Fall 2001, pp. 85–148.

4 United Nations Declaration on Human Cloning, 23 March 2005.

5 Aaron Levine, "Geographic Trends in Human Embryonic Stem Cell Research," *Politics and the Life Sciences*, 23(2), September 2005, pp. 40–45; Jason Owen-Smith and Jennifer McCormick, "An International Gap in Human ES Research," *Nature Biotechnology*, 24(4), April 2006, pp. 391–392.

6 Aaron D. Levine, "Research Policy and the Mobility of US Stem Cell Scientists" *Nature Biotechnology*, 24(7), July 2006, pp. 865–866.

7 "S. Korean Stem-Cell Scientist has 14 Patents: Report," *Asia Pulse*, 16 September 2005.

8 Alexander Meissner and Rudolf Jaenisch, "Generation of Nuclear Transfer-Derived Pluripotent ES Cells from Cloned Cdx2-Deficient Blastocysts," *Nature*, 439(7073), 12 January 2006, pp. 212–215.

9 Irina Klimanskaya *et al.*, "Human Embryonic Stem Cell Lines Derived from Single Blastomeres," *Nature*, 444(7118), 23 November 2006, pp. 481–485.

The future of cloning

1 Pew Initiative on Food and Biotechnology. "Public Sentiment about Genetically Modified Food – November 2005 Update."

2 Liangxue Lai *et al.*, "Generation of Cloned Transgenic Pigs Rich in Omega-3 Fatty Acids," *Nature Biotechnology*, 24(4), April 2006, pp.435–436.

3 Duncan Gardham, "Extremists' protests halt GM crop trials," *Daily Telegraph* (London), 21 March 2005, p. 6.

4 William Saletan "The Organ Factory," *Slate*, 25 July 2005.

5 David Adam, "Faking babies: scientists are developing artificial wombs, sperm and eggs – but will this lead to reproduction in a dish?" *Guardian* (London), 19 May 2005, p. 4.

Glossary

Activation
Also called egg activation. A component of fertilization triggered by sperm penetration. As part of this process, the exterior of the egg changes to prevent additional sperm from penetrating. The egg is also induced to complete its second meiotic division.

Altered nuclear transfer
A proposed alternative to somatic cell nuclear transfer that attempts to side-step the ethical debates associated with deriving human embryonic stem cell lines. In this procedure, either the somatic cell or the enucleated egg to be used in the cloning procedure is altered in such a way that the resulting cloned embryo can develop to the point that embryonic stem cells can be isolated but the embryo cannot develop normally to term.

Asexual reproduction
Reproduction where all genetic material comes from a single parent.

Base
One of the building blocks that make up DNA or RNA. In DNA, the four bases are adenine, cytosine, guanine, and thymine. They are typically abbreviated A, C, G, and T.

Blastocyst
An early stage preimplantation embryo (typically day five to six in human development). At this stage the embryo is a hollow sphere

with two distinct cell types. The outer layer of cells is trophectoderm that will go on to form the placenta, while the inner layer of
cells is the inner cell mass, the cells that will give rise to the mature
organism.

Cell
The basic subunit of living organisms. Some organisms, such as
bacteria, consist of just one cell. Higher organisms, such as
humans, are made up of a huge number of two hundred or so different types of cells.

Chromosome
A threadlike structure, consisting of both proteins and DNA, found
in the nucleus of most cells. The DNA on a chromosome carries
genetic information in the form of genes. In humans, each cell contains 46 chromosomes, 23 from the mother and 23 from the father.

Cleavage
The process of cell division in early development by which the
zygote develops into a blastocyst. This division occurs without
appreciable growth in the size of the embryo, so each division
divides the zygote into smaller and smaller cells.

Clone
An organism produced through asexual reproduction and thus
genetically identical to its single parent

Culture (cell culture)
A technique for growing cells outside the body in laboratory
conditions.

Cytoplasm
The contents of a cell outside the nucleus.

DNA

Short for deoxyribonucleic acid. The material inside cells that carries genetic information. Large molecule consisting of a sequence of bases, or nucleotides.

DNA methylation

An epigenetic modification to DNA in which a small molecule, called a methyl group, is added to DNA. This does not change the coding sequence but can affect the expression of a gene.

Ectoderm

The outermost of the three primary germ layers. Gives rise to the skin and the nervous system.

Embryo

A general term used to describe the early stages of development. In humans, embryo is typically used until the eighth week of pregnancy, after which the developing organism is referred to as a fetus.

Embryonic stem cell

A cell found in the inner cell mass of blastocyst-stage embryos that has the potential to give rise to every cell in the mature organism.

Embryonic stem cell line

When embryonic stem cells are isolated from embryos and grown successfully in cell culture, the resulting cells are called an embryonic stem cell line. Once established as a cell line, these cells can grow and remain undifferentiated almost indefinitely.

Embryoid body

A cluster of partially differentiated cells that arises spontaneously if embryonic stem cells are grown in suspension, rather than on a flat surface. Scientists can examine the cells found in an embryoid body to verify that particular cells are pluripotent.

Endoderm

The innermost of the three primary germ layers. Gives rise to the digestive system and lungs.

Enucleated egg

An unfertilized egg with its nuclear genetic material removed. Eggs are enucleated as the first step in a somatic cell nuclear transfer procedure.

Enzyme

A protein that serves to catalyze a specific chemical reaction within the cell.

Epigenetic changes

Heritable changes in gene function that do not involve changes to the DNA sequence. One of the most common epigenetic changes is DNA methylation.

Fertilization

The multi-step process through which an egg and sperm fuse to form a zygote.

Gastrulation

An important step in the developmental process. Cells from the inner cell mass of a developing embryo fold in on themselves, creating distinct cell layers. This marks the first substantial differentiation of cells that will give rise to the mature organism.

Gene

A sequence of DNA that represents a single unit of heredity. Genes are aligned on chromosomes.

Genetic code

The language in which DNA's instructions are written. The code consists of three-letter blocks of DNA, each of which codes for a specific amino acid. This code is translated by cellular machinery to make proteins.

Genetic determinism
The idea that humans and other animals are completely determined by their genes. Ignores the important role environment plays in development.

Germ cell
A reproductive cell. Either an egg or sperm or one of the cells from which these cells are derived.

Implantation
The attachment of a developing embryo to the uterine wall. This process includes the hatching of the embryo from the *zona pellucida*, which protected it during its passage through the oviducts.

Imprinting
A relatively rare effect where the expression of a gene depends on whether or not it is on a chromosome inherited from the mother or the father.

In vitro
In an artificial environment, such as a test tube or petri dish.

In Vitro fertilization
An assisted reproductive technology. First used successfully in 1978, this technique involves the fertilization of an egg by sperm in a petri dish. The fertilized egg is then allowed to begin development in preparation for transfer to the woman's uterus. Typically many more embryos are created by this process than are needed for reproductive purposes and debates have arisen over whether or not these embryos should be used for human embryonic stem cell research.

In vivo

In the body.

Inner cell mass

A small clump of cells in a blastocyst-stage embryo that will eventually give rise to every cell in the mature organism. It is these cells that are removed from the embryo and grown in culture when scientists derive embryonic stem cell lines.

Interspecies nuclear transfer

A variant of nuclear transfer that crosses species boundaries, typically because the somatic cell donor and the egg donor are of different species. This technique is often attempted because eggs of one species are in short supply. It has been used to try to clone endangered species. The result is an organism with nuclear DNA from one species and mtDNA from another.

Meiosis

The cell division process that gives rise to egg and sperm cells. In this process, DNA replicates once but the cells divide twice, leaving four cells with half the DNA of the original parent cell. Contrast with mitosis.

Mesoderm

Middle layer of the three primary germ layers. Gives rise to bone, muscle, and connective tissue.

Mitochondria

The energy-producing component of cells. Mitochondria are found throughout the cytoplasm of cells and contain a small amount of DNA, called mitochondrial DNA or mtDNA.

Mitosis

The cell division process used by somatic cells. Each daughter cell is genetically identical to its parent cell. Contrast with meiosis.

Morula
An early stage of embryonic development (before the blastocyst) in which the embryo is a small solid mass of cells formed from cleavage of the zygote.

Nuclear envelope
Membrane boundary that separates the nucleus from the rest of the cell.

Nucleotide
One of the building blocks of DNA or RNA. Also called a base.

Nucleus
The component of a cell that contains the genetic material.

Parthenogenesis
Development of an organism from an unfertilized egg. In some lower organisms, this development can lead to mature organisms. In mammals, development halts after a small number of cell divisions.

Placenta
A temporary tissue that connects a developing embryo or fetus to its mother and provides for the transport of oxygen, water, and other nutrients between the two.

Pluripotent
A term used to describe cells that can divide and give rise to many but not necessarily all differentiated cells in an organism.

Primitive streak
A line that appears on the developing embryo during gastrulation. The first visible indicator of differentiation.

Reproductive cloning
The use of somatic cell nuclear transfer to create a new organism genetically identical to its single parent.

Somatic cell
All cells in a multicellular organism with the exception of egg and sperm cells or their precursor cells.

Somatic cell nuclear transfer
The most commonly used technique for cloning animals. A single somatic cell or somatic cell nucleus is transferred to an enucleated egg and the egg is activated so it begins development.

Tetraploid embryo complementation
A method for verifying the pluripotency of an embryonic stem cell line. Embryonic stem cells are injected into specially modified blastocyst stage embryos, which cannot develop on their own. Injection of embryonic stem cells rescues these modified embryos and permits normal development. All cells in resulting organisms are derived from the injected embryonic stem cells, proving the injected cells were pluripotent.

Therapeutic cloning
The use of somatic cell nuclear transfer to create a blastocyst-stage embryo from which embryonic stem cells can be isolated. Also called research cloning.

Totipotent
A term used to describe cells that can divide and give rise to all differentiated cells in an organism, including extra-embryonic tissues.

Transcription factor
Molecule within the cell that exerts control over a cell's gene expression program. Transcription factors present in the egg cell

are thought to play an important role in the reprogramming of a somatic cell nucleus during the cloning procedure.

Transgenic animal

An animal that has been modified to contain a gene or genes from another species.

Trophectoderm

Outer layer of cells at the blastocyst stage that will go on to form the placenta and other extra-embryonic layers.

True-breeding

A line of plants or animals that exhibit the same genetic characteristics generation after generation.

Xenotransplantation

Transplant that crosses species boundaries. For instance, a transplant of a pig organ into a human would be a xenotransplant.

Zona pellucida

Protective covering around the egg and early embryo. Must be penetrated by a sperm before fertilization.

Zygote

Fertilized egg formed from the union of egg and sperm.

Index